W9-BKI-471

## Advance Praise for
# *REBOOTING AI*

"A welcome antidote to the hype that has engulfed AI over the past decade and a realistic look at how far AI and robotics still have to go."

—Rodney Brooks, former director of the MIT Computer Science and Artificial Intelligence Laboratory

"AI is achieving superhuman performance in many narrow applications, but the reality is that we are still very far from artificial general intelligence that truly understands the world. Gary Marcus and Ernest Davis explain the pitfalls of current approaches with humor and insight, and they provide a compelling path toward the kind of robust AI that can earn our trust."

—Erik Brynjolfsson, professor, MIT Sloan School of Management, and co-author of *The Second Machine Age* and *Machine Platform Crowd*

"*Rebooting AI* is a blast to read. It's erudite, it's witty, and it neatly unpacks why today's AI has such trouble doing truly smart tasks—and what it'll take to reach that goal."

—Clive Thompson, *Wired* magazine columnist and author of *Coders: The Making of a New Tribe and the Remaking of the World*

"Will machines overtake humans in the foreseeable future, or is it just hype? Marcus and Davis lay out their answer with elegant prose and a sure quill, drawing the distinction between today's deep-learning-based, narrow, brittle artificial 'intelligence' and the ever-elusive artificial general intelligence. Common sense and trust, which are intrinsically human, emerge as grand challenges for the field. If you plan to read one book to keep up with AI—this is an outstanding choice!"

—Oren Etzioni, CEO of the Allen Institute for Artificial Intelligence and professor of computer science, University of Washington

## ALSO BY GARY MARCUS

Guitar Zero

Kluge

The Birth of the Mind

The Algebraic Mind

## EDITED BY GARY MARCUS

The Future of the Brain

The Norton Psychology Reader

## ALSO BY ERNEST DAVIS

Representations of
Commonsense Knowledge

Linear Algebra and Probability for
Computer Science Applications

# REBOOTING AI

# REBOOTING AI

## BUILDING ARTIFICIAL INTELLIGENCE WE CAN TRUST

Gary Marcus and Ernest Davis

Grateful acknowledgment is made to Houghton Mifflin Harcourt Publishing
Company for permission to reprint an excerpt from "A Little Girl Tugs at the
Tablecloth," from *Monologue of a Dog: New Poems by Wislawa Szymborska,*
translated from the Polish by Stanislaw Baranczak and Clare Cavanagh.


Library of Congress Cataloging-in-Publication Data
Names: Marcus, Gary, author. Davis, Ernest, author.
Title: Rebooting AI : building artificial intelligence we can trust /
Gary Marcus and Ernest Davis.
Description: First edition. New York : Pantheon Books, 2019.
Includes bibliographical references and index.
Identifiers: LCCN 2019005842. ISBN 9781524748258 (hardcover : alk. paper).
ISBN 9781524748265 (ebook).
Subjects: LCSH: Artificial intelligence.
Classification: LCC Q335 .M368 2019 | DDC 006.3--dc23 |
LC record available at lccn.loc.gov/2019005842

www.pantheonbooks.com

Jacket image by Nadia Snopek/Shutterstock
Jacket design by Kelly Blair

Printed in the United States of America
First Edition

2 4 6 8 9 7 5 3 1

For my children, Alexander and Chloe,
who have taught me so much,
and my wife, Athena,
who shares my zest for learning from them.

*Gary*

---

For my wife, Bianca,
the love of my life.

*Ernie*

Although this wave of popularity is certainly pleasant and exciting for those of us working in the field, it carries at the same time an element of danger. While we feel that information theory is indeed a valuable tool in providing fundamental insights into the nature of communication problems and will continue to grow in importance, it is certainly no panacea for the communication engineer or, a fortiori, for anyone else. Seldom do more than a few of nature's secrets give way at one time. It will be all too easy for our somewhat artificial prosperity to collapse overnight when it is realized that the use of a few exciting words like information, entropy, redundancy, do not solve all our problems.

—CLAUDE E. SHANNON, "THE BANDWAGON," *IRE TRANSACTIONS ON INFORMATION THEORY,* 1(2) (1956):3

Any fool can know. The point is to understand.

—ORIGIN UNKNOWN, OFTEN ATTRIBUTED TO ALBERT EINSTEIN

# Contents

# REBOOTING AI

# 1

# Mind the Gap

> Machines will be capable, within twenty years, of doing any work a man can do.
>
> —AI PIONEER HERB SIMON, 1965

> FIRST CHILD [*on a long, arduous journey*]: Is it much further, Papa Smurf?
>
> FATHER: Not far now.
>
> SECOND CHILD [*much later*]: Is it much further, Papa Smurf?
>
> FATHER: Not far now.
>
> —THE SMURFS

Since its earliest days, artificial intelligence has been long on promise, short on delivery. In the 1950s and 1960s, pioneers like Marvin Minsky, John McCarthy, and Herb Simon genuinely believed that AI could be solved before the end of the twentieth century. "Within a generation," Marvin Minsky famously wrote, in 1967, "the problem of artificial intelligence will be substantially solved." Fifty years later, those promises still haven't been fulfilled, but they have never stopped coming. In 2002, the futurist Ray Kurzweil made a public bet that AI would "surpass native human intelligence" by 2029. In November 2018 Ilya Sutskever, co-founder of OpenAI, a major AI research institute, suggested that "near term AGI [artificial general intelligence] should be taken seriously as a possibility." Although it is still theoretically possible that Kurzweil and Sutskever might turn out to be right, the odds against this happening are very long. Getting to that level—general-purpose artificial intelligence with the flexibility of human intelligence—isn't some small step from where we are now;

instead it will require an immense amount of foundational progress—not just more of the same sort of thing that's been accomplished in the last few years, but, as we will show, something entirely different.

Even if not everyone is as bullish as Kurzweil and Sutskever, ambitious promises still remain common, for everything from medicine to driverless cars. More often than not, what is promised doesn't materialize. In 2012, for example, we heard a lot about how we would be seeing "autonomous cars [in] the near future." In 2016, IBM claimed that Watson, the AI system that won at *Jeopardy!,* would "revolutionize healthcare," stating that Watson Health's "cognitive systems [could] understand, reason, learn, and interact" and that "with [recent advances in] cognitive computing . . . we can achieve more than we ever thought possible." IBM aimed to address problems ranging from pharmacology to radiology to cancer diagnosis and treatment, using Watson to read the medical literature and make recommendations that human doctors would miss. At the same time, Geoffrey Hinton, one of AI's most prominent researchers, said that "it is quite obvious we should stop training radiologists."

In 2015 Facebook launched its ambitious and widely covered project known simply as M, a chatbot that was supposed to be able to cater to your every need, from making dinner reservations to planning your next vacation.

As yet, none of this has come to pass. Autonomous vehicles may someday be safe and ubiquitous, and chatbots that can cater to every need may someday become commonplace; so too might superintelligent robotic doctors. But for now, all this remains fantasy, not fact.

The driverless cars that do exist are still primarily restricted to highway situations with human drivers required as a safety backup, because the software is too unreliable. In 2017, John Krafcik, CEO at Waymo, a Google spinoff that has been working on driverless cars for nearly a decade, boasted that Waymo would shortly have driverless cars with no safety drivers. It didn't happen. A year later, as *Wired* put it, the bravado was gone, but the safety drivers weren't. Nobody really thinks that driverless cars are ready to drive fully on their own in cities or in bad weather, and early optimism has been replaced by widespread recognition that we are at least a decade away from that point—and quite possibly more.

IBM Watson's transition to health care similarly has lost steam. In 2017, MD Anderson Cancer Center shelved its oncology collaboration with IBM. More recently it was reported that some of Watson's recommendations were "unsafe and incorrect." A 2016 project to use Watson for the diagnosis of rare diseases at the Marburg, Germany, Center for Rare and Undiagnosed Diseases was shelved less than two years later, because "the performance was unacceptable." In one case, for instance, when told that a patient was suffering from chest pain, the system missed diagnoses that would have been obvious even to a first year medical student, such as heart attack, angina, and torn aorta. Not long after Watson's troubles started to become clear, Facebook's M was quietly canceled, just three years after it was announced.

Despite this history of missed milestones, the rhetoric about AI remains almost messianic. Eric Schmidt, the former CEO of Google, has proclaimed that AI would solve climate change, poverty, war, and cancer. XPRIZE founder Peter Diamandis made similar claims in his book *Abundance,* arguing that strong AI (when it comes) is "definitely going to rocket us up the Abundance pyramid." In early 2018, Google CEO Sundar Pichai claimed that "AI is one of the most important things humanity is working on . . . more profound than . . . electricity or fire." (Less than a year later, Google was forced to admit in a note to investors that products and services "that incorporate or utilize artificial intelligence and machine learning, can raise new or exacerbate existing ethical, technological, legal, and other challenges.")

Others agonize about the potential dangers of AI, often in ways that show a similar disconnect from current reality. One recent nonfiction bestseller by the Oxford philosopher Nick Bostrom grappled with the prospect of superintelligence taking over the world, as if that were a serious threat in the foreseeable future. In the pages of *The Atlantic,* Henry Kissinger speculated that the risk of AI might be so profound that "human history might go the way of the Incas, faced with a Spanish culture incomprehensible and even awe-inspiring to them." Elon Musk has warned that working on AI is "summoning the demon" and a danger "worse than nukes," and the late Stephen Hawking warned that AI could be "the worst event in the history of our civilization."

But what AI, exactly, are they talking about? Back in the real world, current-day robots struggle to turn doorknobs, and Teslas driven in "Autopilot" mode keep rear-ending parked emergency vehicles (at least four times in 2018 alone). It's as if people in the fourteenth century were worrying about traffic accidents, when good hygiene might have been a whole lot more helpful.

---

One reason that people often overestimate what AI can actually do is that media reports often overstate AI's abilities, as if every modest advance represents a paradigm shift.

Consider this pair of headlines describing an alleged breakthrough in machine reading.

> "Robots Can Now Read Better than Humans, Putting Millions of Jobs at Risk"
>
> —*NEWSWEEK*, JANUARY 15, 2018

> "Computers Are Getting Better than Humans at Reading"
>
> —*CNN MONEY*, JANUARY 16, 2018

The first is a more egregious exaggeration than the second, but both wildly oversell minor progress. To begin with, there were no actual robots involved, and the test only measured one tiny aspect of reading. It was far from a thorough test of comprehension. No actual jobs were remotely in jeopardy.

All that happened was this. Two companies, Microsoft and Alibaba, had just built programs that made slight incremental progress on a particular test of a single narrow aspect of reading (82.65 percent versus the previous record of 82.136 percent), known as SQuAD (the Stanford Question Answering Dataset), arguably achieving human-level performance on that specific task where they weren't quite at human level before. One of the companies put out a press release that made this minor achievement sound much more revolutionary than it really was, announcing the creation of "AI that can read a document and answer questions about it as well as a person."

Reality was much less sexy. Computers were shown short passages of text drawn from an exam designed for research purposes and asked questions about them. The catch is that in every case the correct answers appeared *directly in the text*—which rendered the exam an exercise in underlining, and nothing more. Untouched was much of the real challenge of reading: inferring meanings that are implied yet not always fully explicit.

Suppose, for example, that we hand you a piece of paper with this short passage:

> Two children, Chloe and Alexander, went for a walk. They both saw a dog and a tree. Alexander also saw a cat and pointed it out to Chloe. She went to pet the cat.

It is trivial to answer questions like "Who went for a walk?," in which the answer ("Chloe and Alexander") is directly spelled out in the text, but any competent reader should just as easily be able to answer questions that are not directly spelled out, like "Did Chloe see the cat?" and "Were the children frightened by the cat?" If you can't do that, you aren't really following the story. Because SQuAD didn't include any questions of this sort, it wasn't really a strong test of reading; as it turns out the new AI systems would not have been able to cope with them.* By way of contrast, Gary tested the story

* Even easier questions like "What did Alexander see?" would be out of bounds, because the answer (a dog, a tree, and a cat) requires highlighting two pieces of text that aren't contiguous, and the test had made it easy on machines by restricting questions to those that could be answered with a single bit of contiguous text.

on his daughter Chloe, then four and a half years old, and she had no trouble making the inference that the fictitious Chloe had seen a cat. (Her older brother, then not quite six years old, went a step further, musing about what would happen if the dog actually turned out to be a cat; no current AI could begin to do that.)

Practically every time one of the tech titans puts out a press release, we get a reprise of this same phenomenon, in which a minor bit of progress is portrayed in many (mercifully not all) media outlets as a revolution. A couple of years ago, for example, Facebook introduced a bare-bones proof-of-concept program that read simple stories and answered questions about them. A slew of enthusiastic headlines followed, like "Facebook Thinks It Has Found the Secret to Making Bots Less Dumb" (*Slate*) and "Facebook AI Software Learns and Answers Questions. Software able to read a synopsis of Lord of the Rings and answer questions about it could beef up Facebook search" (*Technology Review*).

That really would be a major breakthrough—if it were true. A program that could assimilate even the *Reader's Digest* or Cliffs-Notes versions of Tolkien (let alone the real thing) would be a major advance.

Alas, a program genuinely capable of such a feat is nowhere in sight. The synopsis that the Facebook system actually read was just four lines long:

> Bilbo travelled to the cave. Gollum dropped the ring there. Bilbo took the ring. Bilbo went back to the Shire. Bilbo left the ring there. Frodo got the ring. Frodo journeyed to Mount Doom. Frodo dropped the ring there. Sauron died. Frodo went back to the Shire. Bilbo travelled to the Grey Havens. The End.

And even then, all the program could do was answer basic questions directly addressed in those sentences, such as "Where is the ring?," "Where is Bilbo now?," and "Where is Frodo now?" Forget about asking why Frodo dropped the ring.

The net effect of a tendency of many in the media to overreport technology results is that the public has come to believe that AI is much closer to being solved than it really is.

Whenever you hear about a supposed success in AI, here's a list of six questions you could ask:

1. Stripping away the rhetoric, what did the AI system actually do here?
2. How general is the result? (E.g., does an alleged reading task measure all aspects of reading, or just a tiny slice of it?)
3. Is there a demo where I can try out my own examples? (Be very skeptical if there isn't.)
4. If the researchers (or their press people) allege that an AI system is better than humans, then which humans, and how much better?
5. How far does succeeding at the particular task reported in the new research actually take us toward building genuine AI?
6. How robust is the system? Could it work just as well with other data sets, without massive amounts of retraining? (E.g., could a game-playing machine that mastered chess also play an action-adventure game like Zelda? Could a system for recognizing animals correctly identify a creature it had never seen before as an animal? Would a driverless car system that was trained during the day be able to drive at night, or in the snow, or if there was a detour sign not listed on its map?)

This book is about how to be skeptical, but more than that, it's about why AI, so far, hasn't been on the right track, and what we might do to work toward AI that is robust and reliable, capable of functioning in a complex and ever-changing world, such that we can genuinely trust it with our homes, our parents and children, our medical decisions, and ultimately our lives.

---

To be sure, AI has been getting more impressive, virtually every day, for the last several years, sometimes in ways that are truly amazing.

There have been major advances in everything from game playing to speech recognition to identifying faces. A startup company we are fond of, Zipline, uses a bit of AI to guide drones to deliver blood to patients in Africa, a fantastic application that would have been out of the question just a few years ago.

Much of this recent success in AI has been driven largely by two factors: first, advances in hardware, which allow for more memory and faster computation, often by exploiting many machines working in parallel; second, big data, huge data sets containing gigabytes or terabytes (or more) of data that didn't exist until a few years ago, such as ImageNet, a library of 15 million labeled pictures that has played a pivotal role in training computer vision systems; Wikipedia; and even the vast collections of documents that together make up the World Wide Web.

Emerging in tandem with the data has been an algorithm for churning through that data, called *deep learning,* a kind of powerful statistical engine that we will explain and evaluate in chapter 3. Deep learning has been at the center of practically every advance in AI in the last several years, from DeepMind's superhuman Go and chess player AlphaZero to Google's recent tool for conversation and speech synthesis, Google Duplex. In each case, big data plus deep learning plus faster hardware has been a winning formula.

Deep learning has also been used with substantial success for a wide range of practical applications from diagnosing skin cancer to predicting earthquake aftershocks to detecting credit card fraud. It's also been used in art and music, as well as for a huge number of commercial applications, from deciphering speech to labeling photos to organizing people's news feeds. You can use deep learning to identify plants or to automatically enhance the sky in your photos and even to colorize old black-and-white pictures.

Along with deep learning's stunning success, AI has become a huge business. Companies like Google and Facebook are in an epic battle for talent, often paying PhDs the sort of starting salaries we expect for professional athletes. In 2018, the most important scientific conference for deep learning sold out in twelve minutes. Although we will be arguing that AI with human-level flexibility is much harder than many people think, there is no denying that real progress has been made. It's not an accident that the broader public has become excited about AI.

Nations have, too. Countries like France, Russia, Canada, and China have all made massive commitments to AI. China alone is planning to invest $150 billion by 2030. The McKinsey Global Institute estimates that the overall economic impact of AI could be $13 trillion, comparable to the steam engine in the nineteenth century and information technology in the twenty-first.

Still, that doesn't guarantee that we are on the right path.

---

Indeed, even as data has become more plentiful, and clusters of computers have become faster, and investments bigger, it is important to realize that something fundamental is still missing. Even with all the progress, in many ways machines are still no match for people.

Take reading. When you read or hear a new sentence, your brain, in less than a second, performs two types of analysis: (1) it parses the sentence, deconstructing it into its constituent nouns and verbs and what they mean, individually and collectively; and (2) it connects that sentence to what you know about the world, integrating the grammatical nuts and bolts with a whole universe of entities and ideas. If the sentence is a line of dialogue in a movie, you update your understanding of a character's intentions and prospects. Why did they say what they said? What does that tell us about their character? What are they trying to achieve? Is it truthful or deceptive? How does it relate to what has happened before? How will their speech affect others? For example, when thousands of former slaves stand up one by one and declare "I am Spartacus," each one risking execution, we all know instantly that every one of them (except Spartacus himself) is lying, and that we have just witnessed something moving and profound. As we will demonstrate, current AI programs can't do or understand anything remotely like this; as far as we can tell, they aren't even on track to do so. Most of the progress that has been made has been on problems like object recognition that are entirely different from challenges in understanding meaning.

The difference between the two—object recognition and genuine comprehension—matters in the real world. The AI programs that power the social media platforms we have now, for example, can help spread fake news, by feeding us outrageous stories that garner clicks, but they can't understand the news well enough to judge which stories are fake and which are real.

Even the prosaic act of driving is more complex than most people realize. When you drive a car, 95 percent of what you do is absolutely routine and easily replicated by machines, but the first time a teenager darts out in front of your car on a battery-powered hoverboard, you will have to do something no current machine can do reliably: reason and act on something new and unexpected, based not on some immense database of prior experience, but on a powerful and flexible understanding of the world. (And you can't just slam on the brakes every time you see something unexpected, or you could get rear-ended every time you stop for a pile of leaves in the road.)

Currently, truly driverless cars can't be counted on. Perhaps the closest thing commercially available to consumers is Autopilot-equipped Teslas, but they still demand the full attention of the human driver at all times. The system is reasonably reliable on highways in good weather, but less likely to be reliable in dense urban areas. On a rainy day in the streets of Manhattan or Mumbai, we would still trust our lives sooner to a randomly chosen human driver than to a driverless car.* The technology just isn't mature yet. As a

* Directly comparable data for comparing the safety of humans versus machines have not yet been published. Much of the testing has been done on highways, which are easiest for machines, rather than in crowded urban areas, which pose greater challenges for AI. Published data suggest that the most reliable extant software requires human intervention roughly once every 10,000 miles, even in rather easy driving conditions. By way of imperfect comparison, human drivers are involved in fatal accidents on average

Toyota vice president for automated driving research recently put it, "Taking me from Cambridge to Logan Airport with no driver in any Boston weather or traffic condition—that might not be in my lifetime."

Likewise, when it comes to understanding the plot of a movie or the point of a newspaper article, we would trust middle school students over any AI system. And, much as we hate changing diapers, we can't imagine any robot now in development being reliable enough to help.

o———o

The central problem, in a word: current AI is *narrow;* it works for particular tasks that it is programmed for, provided that what it encounters isn't too different from what it has experienced before. That's fine for a board game like Go—the rules haven't changed in 2,500 years—but less promising in most real-world situations. Taking AI to the next level will require us to invent machines with substantially more flexibility.

What we have for now are basically digital idiots savants: software that can, for example, read bank checks or tag photos or play board games at world champion levels, but does little else. Riffing off a line from investor Peter Thiel about wanting flying cars and instead getting 140 characters, we wanted Rosie the Robot, ready at a moment's notice to change our kids' diapers and whip up dinner, and instead we got Roomba, a hockey-puck-shaped vacuum cleaner with wheels.

Or consider Google Duplex, a system that makes phone calls and sounds remarkably human. When it was announced in spring 2018 there was plenty of discussion about whether computers should be required to identify themselves when making such phone calls.

---

only once in every 100 million miles. One of the greatest risks in driverless cars is that if the machine requires intervention only infrequently, we are apt to tune out, and not be available quickly enough when intervention is required.

(Under much public pressure, Google agreed to that after a couple of days.) But the real story is how narrow Duplex was. For all the fantastic resources of Google (and its parent company, Alphabet), the system that they created was so narrow it could handle just three things: restaurant reservations, hair salon appointments, and the opening hours of a few selected businesses. By the time the demo was publicly released, on Android phones, even the hair salon appointments and the opening hour queries were gone. Some of the world's best minds in AI, using some of the biggest clusters of computers in the world, had produced a special-purpose gadget for making nothing but restaurant reservations. It doesn't get narrower than that.

To be sure, that sort of narrow AI is certainly getting better by leaps and bounds, and undoubtedly there will be more breakthroughs in the years to come. But it's also telling: AI could and should be about so much more than getting your digital assistant to book a restaurant reservation.

It could and should be about curing cancer, figuring out the brain, inventing new materials that allow us to improve agriculture and transportation, and coming up with new ways to address climate change. At DeepMind, now part of Alphabet, there used to be a motto, "Solve intelligence, and then use intelligence to solve everything else." While we think that might have been overpromising a bit—problems are often political rather than purely technical—we agree with the sentiment; progress in AI, if it's large enough, can have major impact. If AI could read and reason as well as humans—yet work with the precision and patience and massive computational resources of modern computer systems—science and technology might accelerate rapidly, with huge implications for medicine and the environment and more. That's what AI should be about. But, as we will show you, we can't get there with narrow AI alone.

Robots, too, could have a much more profound impact than they currently do, if they were powered by a deeper kind of AI than we currently have. Imagine a world in which all-purpose domestic robots have finally arrived, and there are no longer windows to wash, floors to sweep, and, for parents, lunches to pack and diapers to clean. Blind people could use robots as assistants; the elderly could use them as caretakers. Robots could also take over jobs that are dangerous or entirely inaccessible for people, working underground, underwater, in fires, in collapsed buildings, in mine fields, or in malfunctioning nuclear reactors. Workplace fatalities could be greatly reduced, and our capacity to extract precious natural resources might be greatly improved, without putting humans at risk.

Driverless cars, too, could have a profound impact—if we could make them work reliably. Thirty thousand people a year die in the United States in auto accidents, and a million around the globe, and if the AI for guiding autonomous vehicles can be perfected those numbers could be greatly reduced.

The trouble is that the approaches we have now won't take us there, not to domestic robots, or automated scientific discoveries; they probably can't even take us to fully reliable driverless cars. Something important is still missing. Narrow AI alone is not enough.

Yet we are ceding more and more authority to machines that are unreliable and, worse, lack any comprehension of human values. The bitter truth is that for now the vast majority of dollars invested in AI are going toward solutions that are brittle, cryptic, and too unreliable to be used in high-stakes problems.

---

The core problem is trust. The narrow AI systems we have now often work—on what they are programmed for—but they can't be trusted with anything that hasn't been precisely anticipated by their programmers. That is particularly important when the stakes are high. If a narrow AI system offers you the wrong advertisement on Facebook, nobody is going to die. But if an AI system drives your car into an unusual-looking vehicle that isn't in its database, or misdiagnoses a cancer patient, it could be serious, even fatal.

What's missing from AI today—and likely to stay missing, until and unless the field takes a fresh approach—is *broad* (or "general") intelligence. AI needs to be able to deal not only with specific situations for which there is an enormous amount of cheaply obtained relevant data, but also problems that are novel, and variations that have not been seen before.

Broad intelligence, where progress has been much slower, is about being able to adapt flexibly to a world that is fundamentally open-ended—which is the one thing humans have, in spades, that machines haven't yet touched. But that's where the field needs to go, if we are to take AI to the next level.

When a narrow AI plays a game like Go, it deals with a system that is completely *closed;* it consists of a 19-by-19 grid with white and black stones. The rules are fixed and the sheer ability to process many possibilities rapidly puts machines at a natural advantage. The AI can see the entire state of the board and knows all the moves that it and its opponent can legally make. It makes half the moves in the game, and it can predict exactly what the consequences will be. The program can play millions of games and gather a massive amount of trial-and-error data that reflects exactly the environment in which it will be playing.

Real life, by contrast, is open-ended; no data perfectly reflects the ever-changing world. There are no fixed rules, and the possibilities are unlimited. We can't practice every situation in advance, nor foresee what information we will need in any given situation. A system that reads news stories, for example, can't just be trained on what happened last week or last year or even in all recorded history, because new situations pop up all the time. An intelligent news-reading system has to be able to cope with essentially any bit of background information that an average adult might know, even if it's never been critical in a news story before, from "You can use a screwdriver to tighten a screw" to "A chocolate gun is unlikely to be able to fire real bullets." That flexibility is what general intelligence, of the sort any ordinary person has, is all about.

Narrow AI is no substitute. It would be absurd and impractical to have one AI for understanding stories that revolve around tools, and another for those that revolve around chocolate weapons;

there's never going to be enough data to train them all. And no single narrow AI is ever going to get enough data to be able to cover the full range of circumstances. The very act of understanding a story doesn't fit within the whole paradigm of narrow, purely data-driven AI—because the world itself is open.

The open-endedness of the world means that a robot that roamed our homes would similarly confront an essentially infinite world of possibilities, interacting with a vast array of objects, from fireplaces to paintings to garlic presses to internet routers to living beings, such as pets, children, family members, and strangers, and new objects like toys that might only have just gone on the market last week; and the robot must reason about all of them in real time. For example, every painting looks different, but a robot cannot learn separately for each painting what it should and shouldn't do with paintings (leave them on the wall, don't splatter them with spaghetti, and so forth), in some sort of endless mission of trial and error.

Much of the challenge of driving, from an AI perspective, comes from the ways in which driving turns out to be open-ended. Highway driving in good weather is relatively amenable to narrow AI, because highways themselves are largely closed systems; pedestrians aren't allowed, and even cars have limited access. But engineers working on the problem have come to realize that urban driving is much more complex; what can appear on a road in a crowded city at any given moment is essentially unbounded. Human drivers routinely cope with circumstances for which they have little or no direct data (like the first time they see a police officer with a hand-lettered sign saying DETOUR—SINKHOLE). One technical term for such circumstances is that they are *outliers;* narrow AI tends to get flummoxed by them.

Researchers in narrow AI have mostly ignored outliers in a race to build demos and proofs of concept. But the ability to cope with open systems, relying on general intelligence rather than brute force tailored to closed systems, is the key to moving the whole field forward.

This book is about what it would take to make progress toward that more ambitious goal.

It's no exaggeration to say that our future depends on it. AI has enormous potential to help us with some of the largest challenges that face humanity, in key areas including medicine, the environment, and natural resources. But the more power we hand off to AI, the more it becomes critical that AI use that power in ways that we can count on. And that means rethinking the whole paradigm.

---

We call this book *Rebooting AI* because we believe that the current approach isn't on a path to get us to AI that is safe, smart, or reliable. A short-term obsession with narrow AI and the easily achievable "low-hanging fruit" of big data has distracted too much attention away from a longer-term and much more challenging problem that AI needs to solve if it is to progress: the problem of how to endow machines with a deeper understanding of the world. Without that deeper understanding, we will never get to truly trustworthy AI. In the technical lingo, we may be stuck at a local maximum, an approach that is better than anything similar that's been tried, but nowhere good enough to get us where we want to go.

For now, there is an enormous gap—we call it "the AI Chasm"—between ambition and reality.

That chasm has roots in three separate challenges, each of which needs to be faced honestly.

The first we call the *gullibility gap,* which starts with the fact that we humans did not evolve to distinguish between humans and machines—which leaves us easily fooled. We attribute intelligence to computers because we have evolved and lived among human beings who themselves base their actions on abstractions like ideas, beliefs, and desires. The behavior of machines is often superficially similar to the behavior of humans, so we are quick to attribute to machines the same sort of underlying mechanisms, even when they lack them. We can't help but think about machines in cognitive terms ("It thinks I deleted my file"), no matter how simpleminded the rules are that the machines might actually be following. But inferences that are valid when applied to human beings can be entirely off base when applied to AI programs. In homage to a central principle of social psychology, we call this the *fundamental overattribution error.*

One of the first cases of this error happened in the mid-1960s, when a chatbot called Eliza convinced some people that it understood what people were saying to it. In fact, Eliza did little more than match keywords, echo the last thing said, and, when lost, reach for a standard conversational gambit ("Tell me about your childhood"). If you mentioned your mother, it would ask you about your family, even though it had no idea what a family really is, or why one would matter. It was a set of tricks, not a demonstration of genuine intelligence.

Despite Eliza's paper-thin understanding of people, many users were fooled. Some users typed away on a keyboard chatting with Eliza for hours, misinterpreting Eliza's simple tricks for helpful, sympathetic feedback. In the words of Eliza's creator, Joseph Weizenbaum:

> People who knew very well that they were conversing with a machine soon forgot the fact, just as theatergoers, in the grip of suspended disbelief, soon forget that the action they are witnessing is not "real." They would often demand to be permitted to converse with the system in private, and would, after conversing with it for a time, insist, in spite of my explanations, that the machine really understood them.

In other cases, overattribution can literally be deadly. In 2016, a Tesla owner came to trust the Autopilot with his life, to the point that he (allegedly) watched *Harry Potter* while the car chauffeured him around. All was well—until it wasn't. After driving safely for hundreds or thousands of miles, the car literally ran into an unexpected circumstance: a white tractor trailer crossed a highway and the Tesla drove directly underneath the trailer, killing the car's owner. (The car appears to have warned him several times to keep his hands on the wheel, but the driver was presumably too disengaged to respond quickly.) The moral of this story is clear: just because something manages to appear intelligent for a moment or two doesn't mean that it really is, or that it can handle the full range of circumstances a human would.

The second challenge we call the *illusory progress gap:* mistaking progress in AI on easy problems for progress on hard problems. That is what happened with IBM's overpromising about Watson, when progress on *Jeopardy!* was taken to be a bigger step toward understanding language than it really was.

It is possible that DeepMind's AlphaGo could follow a similar path. Go and chess are games of "perfect information"—both players can see the entire board at any moment. In most real-world contexts, nobody knows anything with complete certainty; our data is often noisy and incomplete. Even in the simplest cases, there's plenty of uncertainty; when we decide whether to walk to our doctor's office or take the subway on an overcast day, we don't know exactly how long it will take for the subway to come, or whether it will get stuck, or whether we will be packed in like sardines, or whether we will get soaked if we walk, nor exactly how our doctor will react if we are late. We work with what we've got. Playing Go with yourself a million times, as DeepMind's AlphaGo did, is predictable by comparison. It never had to face uncertainty, or incomplete information, let alone the complexities of human interaction.

There is another way in which games like Go differ deeply from the real world, and it has to do with data: games can be perfectly simulated, so AI systems that play them can collect vast amounts of data cheaply. In Go, a machine can simulate play with humans simply by playing itself; if a system needs billions of data points, it can play itself as often as required. Programmers can get perfectly clean simulation data at essentially no cost. In contrast, in the real world, perfectly clean simulation data doesn't exist, and it's not always possible to collect gigabytes of clean, relevant data by trial and error. In reality, we only get to try out our strategies a handful of times; it's not an option to go to the doctor's office 10 million times, slowly adjusting our parameters with each visit, in order to improve our decisions. If programmers want to train an elder-care robot to help lift infirm people into bed, every data point will cost real money and real human time; there is no way to gather all the data in perfectly reliable simulations. Even crash-test dummies are no substitute. One has to collect data from actual, squirmy people, in different kinds of beds, in different kinds of pajamas, in different kinds of homes, and one can't afford to make mistakes; dropping people a few inches short of the bed would be a disaster. Actual lives are at stake.* As IBM has discovered not once, but twice, first with chess and later with *Jeopardy!*, success in closed-world tasks just doesn't guarantee success in the open-ended world.

The third contributor to the AI Chasm is what we call the *robustness gap*. Time and again we have seen that once people in AI find a solution that works some of the time, they assume that with a little more work (and a little more data) it will work all of the time. And that's just not necessarily so.

Take driverless cars. It's comparatively easy to create a *demo* of a driverless car that keeps to a lane correctly on a quiet road; people have been able to do that for years. It appears to be vastly harder to make them work under circumstances that are challenging or unexpected. As Missy Cummings, director of Duke University's Humans and Autonomy Laboratory (and former U.S. Navy fighter pilot), put it to us in an email, the issue isn't even how many miles a given driverless car might go without an accident, it's how *adaptable* those cars are. In her words, today's semi-autonomous vehicles "typically perform only under extremely narrow conditions which tell you nothing about how they might perform [under] different operating

* Some initial progress has been made here, using narrow AI techniques. AIs have been developed that play more or less as well as the best humans at the video games *Dota 2* and *Starcraft 2,* both of which show only part of the game world to the player at any given moment, and thus involve a form of the "fog of war" challenge. But the systems are narrow and brittle; for example, AlphaStar, which plays *Starcraft 2,* was trained on one particular "race" of character and almost none of its training would carry over to another race. Certainly there is no reason to think that the techniques used in these programs generalize well in complex real-world situations.

environments and conditions." Being almost perfectly reliable across millions of test miles in Phoenix doesn't mean it is going to function well during a monsoon in Bombay.

This confusion—between how autonomous vehicles do in ideal situations (like sunny days on country roads) and what they might do in extreme ones—could well make the difference between success and failure in the entire industry. With so little attention paid to extreme conditions, and so little in the way of methodology to guarantee performance in conditions that are only starting to be examined, it's quite possible that billions of dollars are being wasted on techniques for building driverless cars that simply aren't robust enough to get us to human-grade reliability. We may need altogether different techniques to achieve the final bit of reliability we require.

And cars are just one example. By and large, in contemporary research in AI, robustness has been underemphasized, in part because most current AI effort goes into problems that have a high tolerance for error, such as ad recommendation and product recommendation. If we recommend five products to you and you like only three of them, no harm done. But in many of the most important

A GROUP OF YOUNG PEOPLE PLAYING A GAME OF FRISBEE

*A plausible caption, generated automatically by AI*

potential future applications of AI, including driverless cars, elder care, and medical treatment planning, robustness will be critical. Nobody will buy a home robot that carries their grandfather safely into bed four times out of five.

Even in tasks that are putatively right in the sweet spot of what contemporary AI is supposed to have mastered, trouble looms. Take the challenge of having computers identify what is happening in an image. Sometimes it works, but it frequently doesn't, and the errors are often outlandish. If you show a so-called captioning system a picture of everyday scenes you often get a remarkably human-like answer, as in this scene containing a group of people playing Frisbee (opposite), correctly labeled by a highly touted Google captioning system.

But five minutes later you may get an answer that's entirely absurd, as with this parking sign with stickers on it, mislabeled by the system as a "refrigerator filled with lots of food and drinks."

A REFRIGERATOR FILLED WITH LOTS OF FOOD AND DRINKS

*A less plausible caption, generated by the same system**

* No explanation for the error was given, but such errors are not uncommon. As best we can tell, the system in this specific instance determined

Similarly, driverless cars often correctly identify what they see, but sometimes they don't, as with the Teslas repeatedly crashing into parked fire engines. Similar blind spots in systems controlling power grids or monitoring public health could be even more dangerous.

o———o

To move past the AI Chasm, we need three things: a clear sense of what is at stake, a clear understanding of why current systems aren't getting the job done, and a new strategy.

With so much at stake, in terms of jobs, safety, and the fabric of society, there is an urgent need for lay readers and policy makers alike to understand the true state of the art, and for all of us to learn how to think critically about AI. Just as it is important for informed citizens to understand how easy it is to mislead people with statistics, it is also increasingly important that we be able to sort out AI hype and AI reality, and to understand what AI currently can and cannot do.

Crucially, AI is not magic, but rather just a set of engineering techniques and algorithms, each with its own strengths and weaknesses, suitable for some problems but not others. One of the main reasons we wrote this book is because so much of what we read about AI strikes us as pure fantasy, predicated on a confidence in AI's imagined strengths that bears no relation to current technological capabilities. To a large extent, public discussion about AI has been unmoored from any sort of understanding of the reality of how difficult broad AI would be to achieve.

To be clear: although clarifying all this will require us to be critical, we don't hate AI, we love it. We have spent our professional lives immersed in it, and we want to see it advance, as rapidly as

---

(perhaps in terms of color and texture) that the picture was similar to other pictures that would elicit the label "refrigerator filled with lots of food and drinks"—without realizing (as a human would) that such a description would be appropriate only when there was a large rectangular metal box with various kinds of things inside.

possible. Hubert Dreyfus once wrote a book about what he thought AI couldn't do—ever. Our book isn't like that. It's partly about what AI can't do now—and why that matters—but it's also about what we might do to improve a field that is still struggling. We don't want AI to disappear; we want to see it improve, radically, such that we can truly count on it to solve our problems. We do have plenty of challenging things to say about the current state of AI, but our criticism is tough love, not a call for anyone to give up.

In short, it is our belief that AI truly can transform the world in important ways, but also that many basic assumptions need to change before real progress can be made. *Rebooting AI* is not an argument to shut the field down (though some may read it that way), but rather a diagnosis of where we are stuck—and a prescription for how we might do better.

The best way forward, we will suggest, may be to look inward, toward the structure of our own minds. Truly intelligent machines need not be exact replicas of human beings, but anyone who looks honestly at AI will see that AI still has a lot to learn from people, particularly from small children, who in many ways far outstrip machines in their capacity to absorb and understand new concepts. Pundits often write about computers being "superhuman" in one respect or another, but in five fundamental ways, our human brains still vastly outperform our silicon counterparts: we can understand language, we can understand the world, we can adapt flexibly to new circumstances, we can learn new things quickly (even without gobs of data), and we can reason in the face of incomplete and even inconsistent information. On all of these fronts, current AI systems are non-starters. We will also suggest that a current obsession with building "blank slate" machines that learn everything from scratch, driven purely from data rather than knowledge, is a serious error.

If we want machines to reason, understand language, and comprehend the world, learning efficiently, and with human-like flexibility, we may well need to first understand how humans manage to do so, and to better understand what it is that our minds are even trying to do (hint: it's not all about the kind of correlation-seeking that deep learning excels at). Perhaps it is only then, by meeting these challenges head-on, that we can get the reboot that AI so desperately needs, and create AI systems that are deep, reliable, and trustworthy.

In a world in which AI will soon be as ubiquitous as electricity, nothing could be more important.

# 2

# What's at Stake

> A lot can go wrong when we put blind faith in big data.
>
> —CATHY O'NEIL, TED TALK, 2017

On March 23, 2016, Microsoft released Tay, designed to be an exciting and new chatbot, not hand-wired entirely in advance, like the original chatbot, Eliza, but instead developed largely by learning from user interactions. An earlier project, Xiaoice, which chats in Chinese, had been a huge success in China, and Microsoft had high hopes.

Less than a day later the project was canceled. A nasty group of users tried to drown Tay in racist, sexist, and anti-Semitic hate. Vile speech in, vile speech out; poor Tay was posting tweets like "I fucking hate the feminists" and "Hitler was right: I hate the Jews."

Elsewhere on the internet, there are all sorts of problems great and small. One can read about Alexas that spooked their owners with random giggles, iPhone face-recognition systems that confused mother and son, and the Poopocalypse, the Jackson Pollack–like mess that has happened more than once when a Roomba collided with dog waste.

More seriously, there are hate speech detectors that get easily fooled, job candidate systems that perpetuate bias, and web browsers and recommendation engines powered by AI tools that have been tricked into pushing people toward ludicrous conspiracy theories. In China, a face-recognition system used by police sent a jaywalking ticket to an innocent person who happened to be a well-known entrepreneur when it saw her picture on the side of a bus, not realizing that a larger-than-life photo on a moving bus was not the same thing as the entrepreneur herself. A Tesla, apparently in "Summon" mode, crashed while backing out of its owners' garage. And more than once, robotic lawnmowers have maimed or slaughtered hedgehogs. The AI that we have now simply can't be trusted. Although it often does the right thing, we can never know when it is going to surprise us with errors that are nonsensical or even dangerous.

And the more authority we give them, the more worried we should be. Some glitches are mild, like an Alexa that randomly giggles (or wakes you in the middle of the night, as happened to one of us) or an iPhone that autocorrects what was meant as "Happy Birthday, dear Theodore" into "Happy Birthday, dead Theodore." But others—like algorithms that promote fake news or bias against job applicants—can be serious problems. A report from the group AI Now has detailed many such issues in AI systems in a wide variety of applications, including Medicaid eligibility determination, jail term sentencing, and teacher evaluation. Flash crashes on Wall Street have caused temporary stock market drops, and there have been frightening privacy invasions (like the time an Alexa recorded a conversation and inadvertently sent it to a random person on the owner's contact list); and multiple automobile crashes, some fatal. We wouldn't be surprised to see a major AI-driven malfunction in an electrical grid. If this occurs in the heat of summer or the dead of winter, a large number of people could die.

Which is not to say that we should stay up nights worrying about a Skynet-like world in which robots face off against people, at least any time in the foreseeable future. Robots don't yet have the intelligence or manual dexterity to reliably navigate the world, except in carefully controlled environments. Because their cognitive faculties are so narrow and limited, there is no end to the ways they can be stymied.

More important, there is no reason to think that the robots will, science fiction style, rise up against us. After sixty years of AI, there is not the slightest hint of malice; machines have demonstrated zero interest in tangling with humans for territory, possessions, bragging rights, or anything else that battles have been fought over. They aren't filled with testosterone or an unbridled lust for world domination. Instead, AIs are nerds and idiots savants focused so tightly on what they do that they are unaware of the larger picture.

Take the game of Go, a game that nominally revolves around seizing territory, which is about as close to taking over the world as any extant AI gets. In the 1970s, computer Go programs were terrible, easily beaten by any decent human player, but they showed no signs whatsoever of wishing to meddle with humanity. Forty years later, programs like AlphaGo are fantastic, vastly improved and far better than the best human players; but they still show *zero* interest in seizing human territory, or in relegating their human programmers to a zoo. If it's not on the board, they are not interested.

AlphaGo simply doesn't care about questions like "Is there life outside the Go board?," let alone "Is it fair that my human masters leave me to do nothing but play Go all day?" AlphaGo literally has no life or curiosity at all beyond the board; it doesn't know that the game is usually played with stones, or even that anything exists beyond that grid that it plays on. It doesn't know that it is a computer and uses electric power or that its opponent is a human being. It doesn't remember that it has played many games in the past or foresee that it will be playing more games in the future. It isn't pleased when it wins, nor distressed when it loses, nor proud of the progress it has made in learning to play Go. The sorts of human motives that drive real-world aggression are utterly absent. If you wanted to personify the algorithm (if that even makes sense at all), you would say that AlphaGo is perfectly content doing what it is doing, with zero desire to do anything else.

One could say the same for AI that does medical diagnosis, advertising recommendations, navigation, or anything else. Machines, at least in their current implementation, work on the jobs that they have been programmed to work on, and nothing else. As long as we keep things that way, our worries shouldn't center around some kind of imaginary malice.

As Steven Pinker wrote:

> [T]he scenario [that the robots will become superintelligent and enslave humans] makes about as much sense as the worry that since jet planes have surpassed the flying ability of eagles, someday they will swoop out of the sky and seize our cattle. The . . . fallacy is a confusion of intelligence with motivation—of beliefs with desires, inferences with goals, thinking with wanting. Even if we did invent superhumanly intelligent robots, why would they *want* to enslave their masters or take over the world? Intelligence is the ability to deploy novel means to attain a goal. But the goals are extraneous to the intelligence: Being smart is not the same as wanting something.

To take over the world, the robots would have to want to; they'd have to be aggressive, ambitious, and unsatisfied, with a violent streak. We've yet to encounter a robot remotely like that. And for now there is no reason to build robots with emotional states at all, and no compelling idea about how we could do so even if we wanted to. Humans may use emotions like discontent as a tool for motivation, but robots don't need anything of the kind to show up for work; they just do what they are told.

We don't doubt that robots might someday have physical and intellectual powers that could potentially make them formidable foes—*if* they chose to oppose us—but at least for the foreseeable future, we don't see any reason that they would.

Still, we are not home free. AI doesn't have to *want* to destroy us in order to create havoc. In the short term, what we should worry most about is whether machines are actually capable of *reliably* doing the tasks that we assign them to do.

A digital assistant that schedules our appointments is invaluable—if it's reliable. If it accidentally sends us to a critical meeting a week late, it's a disaster. And even more will be at stake when we make the inevitable transition to home robots. If some corporate titan designs a home robot to make crème brûlée, we want it to work every time, not just nine times out of ten, setting the kitchen on fire on the tenth time.

Machines don't, so far as we know, ever have imperialistic ambitions, but they do make mistakes, and the more we rely on them, the more their mistakes matter.

Another problem, for now utterly unsolved, is that machines have to correctly infer our intentions, even when we may be far from explicit, or even manifestly unclear. One issue is what we might call the Amelia Bedelia problem, after the housekeeper in a series of children's stories who takes her employer's requests too literally. Imagine telling your cleaning robot, as you set out in the morning, "Take everything left in the living room and put it away in the closet," only to come back to see *everything*—the TV, the furniture, the carpet—broken into little pieces to fit.

*"Take everything left in the living room and put it away in the closet."*

Then there is the problem of speech errors, particularly likely to happen in caring for elders with cognitive difficulties. If Grandpa asks to have the dinner put in the garbage rather than on the dining table, a good robot should have the sense to make sure that's really what he wants, and not a mistake.

Repurposing a recently popularized phrase, we want our robots and AI to take us seriously, but not always literally.

Of course, all technology can fail, even the oldest and best understood. Not long before we started working on this book, a pedestrian walkway in Miami spontaneously collapsed, killing six people, just five days after it was installed, despite the fact that people have been building bridges for more than three millennia (the Arkadiko Bridge, built in 1300 BC, is still standing).

We can't expect AI to be perfect from day one, and there is a good argument in some cases for tolerating short-term risks in order to achieve long-term gains; if a few people died in the development of driverless cars now, but hundreds of thousands or millions were ultimately saved, the risks might well be worth taking.

That said, until AI is reframed and improved in fundamental ways, risks abound. Here are nine that we worry about the most.

First, there is the fundamental overattribution error we met in the first chapter. AI frequently tempts us into believing that it has a human-like intelligence, even when it doesn't. As the author and MIT social scientist Sherry Turkle has pointed out, a friendly seeming companion robot isn't actually your friend. And we can be too quick to cede power to AI, assuming that success in some context assures reliability in another. One of the most obvious examples of this we already mentioned: the case of driverless cars—good performance in ordinary circumstances is no guarantee of safety in all circumstances. To take a more subtle example, not long ago police officers in Kansas stopped a driver and used Google Translate to get the driver's consent to search his car. A judge later found that the quality of the translation was so poor that the driver could not be considered to have given informed consent, and thus found the search to be in violation of the Fourth Amendment. Until AI gets radically better, we need to be careful in trusting it too much.

Second, there is the lack of robustness. The need for driverless cars to cope with unusual lighting, unusual weather, unusual debris on the road, unusual traffic patterns, humans making unusual gestures, and so forth is one example. Robustness would similarly be essential for a system that truly took charge of your calendar; if it gets confused when you travel from California to Boston and you show up three hours late for a meeting, you have a problem. The need for a better approach to AI is clear.

Third, modern machine learning depends heavily on the precise details of large training sets, and such systems often break if they are applied to new problems that extend beyond the particular data sets on which they were trained. Machine-translation systems trained on legal documents do poorly when applied to medical articles and vice versa. Voice-recognition systems trained only on adult native speakers often have trouble with accents. Technology much like what underlies Tay worked fine when it took its inputs from a society in which political speech is heavily regulated, but produced unacceptable results when it was drowned in a sea of invectives. A deep learning system that could recognize digits when printed in black on a white background with 99 percent accuracy flamed out when the colors were reversed, suddenly getting only 34 percent right—hardly comforting when you stop to reflect on the fact that there are blue stop signs in Hawaii. Stanford computer scientist Judy Hoffman has shown that an autonomous vehicle whose vision system was trained in one city can do significantly worse in another, even in terms of recognizing basic objects like roads, street signs, and other cars.

Fourth, blind data dredging can lead to the perpetuation of obsolete social biases when a more subtle approach is called for. One of the first signs of this came in 2013, when Harvard computer scientist Latanya Sweeney discovered that if you did a Google search on a characteristically black name like Jermaine, you tended to get noticeably more ads offering information about arrest records than if you searched for a characteristically white name like Geoffrey. Then, in 2015, Google Photos mislabeled some photographs of African Americans as gorillas. In 2016 someone discovered that if you did a Google image search for "Professional hair style for work," the pictures returned were almost all of white women, whereas if you searched for "Unprofessional hair style for work," the pictures were almost all of black women. In 2018, Joy Buolamwini, then a graduate student at the MIT Media Lab, found that a raft of commercial algorithms tended to misidentify the gender of African-American women. IBM was the first to patch that particular problem, and Microsoft swiftly followed suit—but so far as we know nobody has come up with anything like a general solution.

Even now, as we write this, it is easy to find similar examples. When we did a Google image search for "mother," the overwhelming majority of the images were white people, an artifact of how data is collected on the web, and an obvious misrepresentation of reality. When we searched for "professor," only about 10 percent of the top-ranked images were women, perhaps reflecting Hollywood's portrayal of college life, but out of touch with current reality, in which closer to half of professors are women. An AI-powered recruiting system that Amazon launched in 2014 was so problematic that they had to abandon it in 2018.

We don't think these problems are insuperable—as we will discuss later, a paradigm shift in AI could help here—but no general solution yet exists.

The core issue is that current AI systems mimic input data, without regard either to social values or to the quality or nature of the data. U.S. government stats tell us that nowadays only 41 percent of faculty are white males, but Google image search doesn't know that; it just moshes together every picture it finds, without a thought about the quality and representativeness of the data or the values that are implicitly expressed. The demographics of faculty are changing, but blind data dredgers miss that, and entrench history rather than reflect changing realities.

Similar concerns arise when we think about the role that AI is starting to play in medicine. The data sets used for training skin cancer diagnostics programs, for example, may be tuned toward white patients, and give invalid results when used on darker-skinned patients. Self-driving cars may be measurably less reliable in recognizing dark-skinned pedestrians than light-skinned ones. Lives are at stake, and current systems are not prepared to grapple with these biases.

Fifth, the heavy dependence of contemporary AI on training sets can also lead to a pernicious echo-chamber effect, in which a system ends up being trained on data that it generated itself earlier. For example, as we will discuss in chapter 4, translation programs work by learning from "bitexts," pairs of documents that are translations of one another. Unfortunately, there are languages where a significant fraction of the texts on the web—in some cases, as much as 50 percent of all web documents—was in fact created by a machine-translation program. As a result, if Google Translate makes some mistake in translation, that mistake can wind up in a document on the web, and that document then becomes data, reinforcing the error.

Similarly, many systems rely on human crowd workers to label images, but sometimes crowd workers use AI-powered bots to do their work for them. Although the AI research community has in turn developed techniques to test whether the work is being done by humans or bots, the whole process has become a cat-and-mouse game between the AI researchers on one side and the mischievous crowd bots on the other, with neither side maintaining a permanent advantage. The result is that a lot of allegedly high-quality human-labeled data turns out to be machine-generated.

Sixth, programs that rely on data that the public can manipulate are often susceptible to being gamed. Tay is of course an example of this. And Google is hit regularly with "Google bombs," in which people create a large number of posts and links so that searches on specific terms give results they find amusing. In July 2018, for example, people succeeded in getting Google Images to respond to searches for "idiot" with pictures of Donald Trump. (This was still true later that year when Sundar Pichai spoke to Congress.) Sixteen years earlier there had been another spoof, rather more indecent, of Rick Santorum. And people don't just game Google for giggles; an entire industry, search engine optimization, exists around manipulating Google into giving a high ranking to its clients in relevant web searches.

Seventh, the combination of preexisting societal biases and the echo effect can lead to the amplification of social bias. Suppose that historically in some city, policing, criminal convictions, and sentencing have been unfairly biased against a particular minority group. The city now decides to use a big data program to advise it on policing and sentencing, and the program is trained on historical data, in which dangerous criminals are identified in terms of their arrest record and jail time. The program will see that dangerous criminals, so defined, disproportionately come from the minority; it will therefore recommend that neighborhoods with a higher percentage of those minorities get more police, and that members of the minority should be more quickly arrested and given longer sentences. When the program is rerun on the new crop of data, the new data reinforces its previous judgments and the program will tend to make the same kinds of biased recommendations with even more confidence.

As Cathy O'Neil, author of *Weapons of Math Destruction,* has emphasized, even if a program is written so as to avoid using race and ethnicity as criteria, there are all kinds of "proxies"—associated features—that it could use instead that would have the same outcome: neighborhood, social media connections, education, jobs, language, even perhaps things like preference in clothing. Moreover, the decisions that the program is making, being computed "algorithmically," have an aura of objectivity that impresses bureaucrats and company executives and cows the general public. The workings of the programs are mysterious—the training data is confidential, the program is proprietary, the decision-making process is a "black box" that even the program designers cannot explain—so it becomes almost impossible for individuals to challenge decisions that they feel are unjust.

Some years ago, Xerox wanted to cut down on costly "churn" among employees, so they deployed a big data program to predict how long an employee would last. The program found that one highly predictive variable was the length of the commute; not surprisingly, employees with long commutes tend to quit sooner. However, the management at Xerox realized that not hiring people with long commutes would in effect amount to discriminating against low- or moderate-income people, since the company was located in a wealthy neighborhood. To its credit, the company removed this as a criterion under consideration. But without close human monitoring, these kinds of biases will surely continue to pop up.

An eighth challenge for AI is that it is all too easy for an AI to end up with the wrong goals. DeepMind researcher Victoria Krakovna has collected dozens of examples of this happening. A soccer-playing robot, encouraged to try to touch the ball as many times as it could, developed a strategy of standing next to the ball and rapidly vibrating—not quite what the programmers had in mind. A robot that was supposed to learn how to grasp a particular object was trained on images of what it looks like to grasp that object, so it decided that it was good enough just to put its hand between the

*Encouraged to try to touch the ball as many times as it can, the robot develops a strategy of standing next to the ball and rapidly vibrating.*

camera and the object, so that it looked like the robot was grasping the object. An unambitious AI tasked with playing *Tetris* decided it was better to pause the game indefinitely rather than risk losing.

The problem of mismatched goals can also take subtler forms. In the early days of machine learning, a dairy company hired a machine-learning company to build a system that could predict when cows are going into estrus. The specified goal for the program was to generate predictions "estrus/no estrus" as accurately as possible. The farmers were delighted to learn that the system was 95 percent accurate. They were less pleased when they found out how the program was managing that. Cows are in estrus for only one day in a twenty-day cycle; based on that, the program's prediction for every single day was the same (no estrus), which made the program correct nineteen days out of twenty—and utterly useless. Unless we spell things out in considerable detail, the solution that we want may not be the one that an AI system alights on.

Finally, because of the scale at which current AI can operate, there are many ways in which AI could (even in its still-primitive form) be used deliberately to cause serious public harm. Stalkers have begun using relatively basic AI techniques to monitor and manipulate their victims, and spammers have used AI for years, to identify potential marks, to evade those CAPTCHAs on websites that make sure you're a human, and so forth. There is little doubt that AI will soon find a role in autonomous weapon systems, though we hold out a modest hope that such techniques might be banned, much as chemical weapons are. As the SUNY political scientist Virginia Eubanks has pointed out, "When a very efficient technology is deployed against a despised outgroup in the absence of strong human rights protections, there is enormous potential for atrocity."

---

None of this means that AI can't do better—but only if there is a fundamental paradigm shift, of the sort we are calling for in this book. We are confident that many of these technical problems can be solved—but not current techniques. Just because contemporary

AI is driven slavishly by the data without any real understanding of the ethical values programmers and system designers might wish to follow doesn't mean that all AI in the future has to be vulnerable to the same problems. Humans look at data, too, but we are not going to decide that almost all fathers and daughters are white, or that the job of a soccer player who has been encouraged to touch the ball more often is to stand next to the ball and vibrate. If humans can avoid these mistakes, machines should be able to, too.

It's not that it is impossible *in principle* to build a physical device that can drive in the snow or manage to be ethical; it's that we can't get there with big data alone.

o———o

What we really need is a new approach altogether, with much more sophistication about what we want in the first place: a fair and safe world. What we see instead are AI techniques that solve individual, narrow problems, while side-stepping the core problems they are intended to address; we have Band-Aids when we need a brain transplant.

IBM, for example, managed to fix the problem of poor gender identification that Joy Buolamwini discovered by building a new training set with more pictures of black women. Google solved its gorilla challenge in the opposite way: by removing pictures of gorillas from the training set. Neither solution is general; both are instead hacks, designed to get blind data analysis to do the right thing, without fixing the underlying problems.

Likewise, one may be able to solve Tesla's problems with running into emergency vehicles stopped on highways by adding better sensors and getting an appropriately labeled set of examples, but who is to say that will work with tow trucks that happened to stop on the side of the highway? Or construction vehicles? Google could hack a fix for the problem of the images of "mother" being almost all white, but then the problem just arises again with "grandmother."

But as long as the dominant approach is focused on narrow AI and bigger and bigger sets of data the field may be stuck playing whack-a-mole indefinitely, finding short-term data patches for particular problems without ever really addressing the underlying flaws that make these problems so common.

What we need are systems smart enough to avoid these errors in the first place.

Today, nearly everyone seems to be pinning their hopes for that on deep learning. As we will explain in the next chapter, we think that's a mistake.

3

# Deep Learning, and Beyond

As to ideas, entities, abstractions, and transcendentals, I could never drive the least conception into their heads.

—JONATHAN SWIFT, *GULLIVER'S TRAVELS*

It's one thing to expect elementary particles to obey simple, universal laws. It's another thing entirely to expect the same of the human race.

—SABINE HOSSENFELDER, *LOST IN MATH*

A large fraction of the current enthusiasm for AI stems from a simple fact: other things being equal, the more data you have, the better. If you want to predict the outcome of the next election, and you can only poll 100 people, good luck; if you can interview 10,000, your chances are much better.

In fact, in the early days of AI, there wasn't much data, and data wasn't a major part of the picture. Most research followed a "knowledge-based" approach, sometimes called GOFAI—Good Old Fashioned AI, or "classical AI." In classical AI, researchers would typically encode by hand the knowledge the AI would need to carry out a particular task, and then write computer programs that leveraged that knowledge, applying it to various cognitive challenges, like understanding stories or making plans for robots or proving theorems. Big data didn't exist, and those systems rarely centered around leveraging data in the first place.

Although it was feasible (often with a great deal of effort) to build laboratory prototypes using this approach, it was often too difficult to get past this stage. The total number of classical AI systems of any practical importance is small. Such techniques are still widely used in certain areas, such as planning routes for robots and GPS-based navigation. Overall, though, the classical, knowledge-focused approach has largely been displaced by *machine learning*, which typically tries to learn everything from data, rather than relying on purpose-built computer programs that leverage hand-coded knowledge.

That approach, too, actually goes back to the 1950s, when Frank Rosenblatt built a "neural network"*—one of the first machine-learning systems—that aimed to learn to recognize objects around it without requiring programmers to anticipate every contingency in advance. His systems got an enormous initial burst of hype and were reported in *The New York Times* to great acclaim in 1958, but these too soon faded, beset by their own problems. His networks (which had to work on 1950s hardware) were underpowered: in today's terminology, they weren't deep enough (we will explain what exactly that means in a moment). His camera also didn't have enough pixels; at 20x20 it had 400, about 1/30,000th of the resolution of an iPhone X, which made for some very pixelated pictures. In retrospect, Rosenblatt had a good idea, but the actual systems he could feasibly build back then just weren't able to do much.

But hardware was only part of the problem. In hindsight, machine learning also depends heavily on having large amounts of data, like pictures named with labels, and Rosenblatt didn't have much; there was no internet from which he could pull millions of examples.

* The term *neural network*, applied to Rosenblatt's devices or to the more complex deep learning systems we will describe later in this chapter, reflects an idea that the components of these devices resemble neurons (nerve cells) in their workings. Some people are attracted to such systems because of their alleged biological plausibility. We think this is a red herring. As we will discuss later, deep learning systems in no way capture the complexity and diversity of actual brains and the components of deep learning systems lack virtually all the complexity of actual neurons. As the late Francis Crick noted, it's a serious stretch to call them brain-like.

Still, many people continued in Rosenblatt's tradition for decades. And until recently, his successors too struggled mightily. Until big data became commonplace, the general consensus in the AI community was that the so-called neural-network approach was hopeless. Systems just didn't work that well, compared to other methods.

When the big data revolution came, in the early 2010s, neural networks finally had their day. People like Geoff Hinton, Yoshua Bengio, Yann LeCun, and Jürgen Schmidhuber, who had stuck by neural networks, even in the dark days of the 1990s and 2000s when most of their colleagues had turned elsewhere, pounced.

In some ways, the most important advance came not from some technical breakthrough in the mathematics of neural networks—much of which had been worked out in the 1980s—but from computer games or, more specifically, from a special piece of hardware known as a GPU (short for graphics processing unit), which the neural-network community repurposed for AI. GPUs had originally been developed for video games, starting in the 1970s, and had been applied to neural networks since the early 2000s. By 2012 they had become extremely powerful, and for certain purposes they were more efficient than CPUs, the traditional core of most computers. A revolution came in 2012, when a number of people, including a team of researchers working with Hinton, worked out a way to use the power of GPUs to enormously increase the power of neural networks.

Suddenly, for the first time, Hinton's team and others began setting records, most notably in recognizing images in the ImageNet database we mentioned earlier. Competitors Hinton and others focused on a subset of the database—1.4 million images, drawn from one thousand categories. Each team trained its system on about 1.25 million of those, leaving 150,000 for testing. Before then, with older machine-learning techniques, a score of 75 percent correct was a good result; Hinton's team scored 84 percent correct, using a deep neural network, and other teams soon did even better; by 2017, image labeling scores, driven by deep learning, reached 98 percent.

Key to this newfound success was the fact that GPUs allowed Hinton and others to "train" neural networks that were much deeper—in the technical sense of having more *layers* (sets of vaguely neuron-like elements we will explain in a moment)—than had been feasible previously. To *train* a deep network is to give the network a bunch of examples, along with correct labels for those examples: this picture is a dog, that picture is a cat, and so forth, a paradigm known as *supervised learning*. Leveraging GPUs meant that more layers could be trained faster, and the results got better.

Between GPUs and the size of the ImageNet library, deep learning was off to the races. Not long after that, Hinton and some grad students formed a company and auctioned it off. Google was the high bidder, and it also bought a startup called DeepMind for more than $500 million two years later. The deep learning revolution began.

o——o

Deep learning itself is ultimately just one approach among many to the challenge of trying to get machines to learn things from data, most often by statistical means.

Say you are running an online bookstore and want to recommend products to your customers. One approach would be to decide, by hand, what your favorite books are. You could put them on your front page, much as physical booksellers put favorite books at the front of the bookstore. But another approach is to learn what people like from data, and not just what people in general like, but what might work for a particular customer based on what they have bought before. You might notice that people who like the *Harry Potter* books also often buy *The Hobbit,* and that people who like Tolstoy often buy Dostoyevsky. As your inventory grows, the number of possibilities expands, and it becomes too hard to keep track of everything individually, so you write a computer program to track it all.

What you are tracking are statistics: the probability that a customer who buys book 1 will also buy book 2, book 3, and so forth. Once you get sophisticated about it, you start tracking more complicated probabilities, like the probability that someone who bought both a *Harry Potter* novel and *The Hobbit* but not a *Star Wars* book will then also buy a science fiction novel by Robert Heinlein. That art of making educated guesses based on data is the large and thriving subfield of AI known as *machine learning.*

A handy way to think about the relation between deep learning, machine learning, and AI is this Venn diagram. AI includes machine learning, but also includes, for example, any necessary algorithm or knowledge that is hand-coded or built by traditional programming techniques rather than learned. Machine learning includes any technique that allows a machine to learn from data; deep learning is the best-known of those techniques, but not the only one.

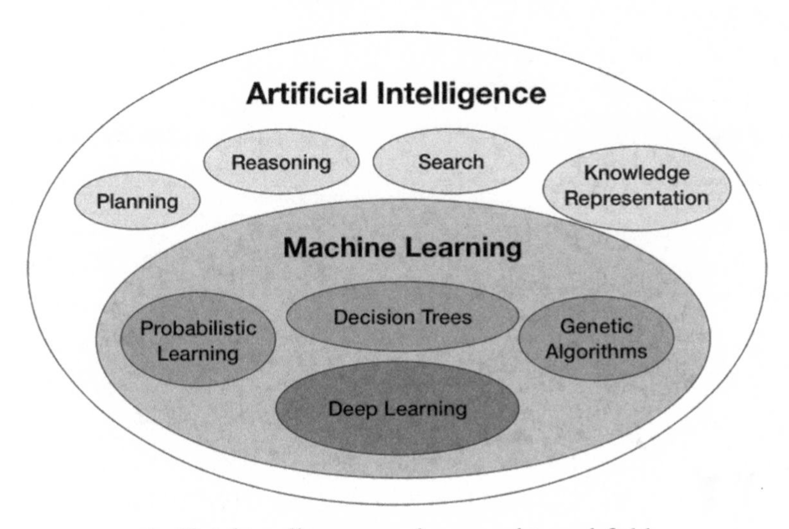


*Artificial intelligence and some of its subfields*

We focus on deep learning here because it is, by a wide margin, the central focus of most current investment in AI, both in the academy and in industry. Still, deep learning is hardly the only approach, either to machine learning or to AI in general. For example, one approach to machine learning is to construct decision trees, which are essentially systems of simple rules that characterize the data.

The support vector machine, a technique that organizes data into complex, abstract hypercubes, was dominant throughout machine

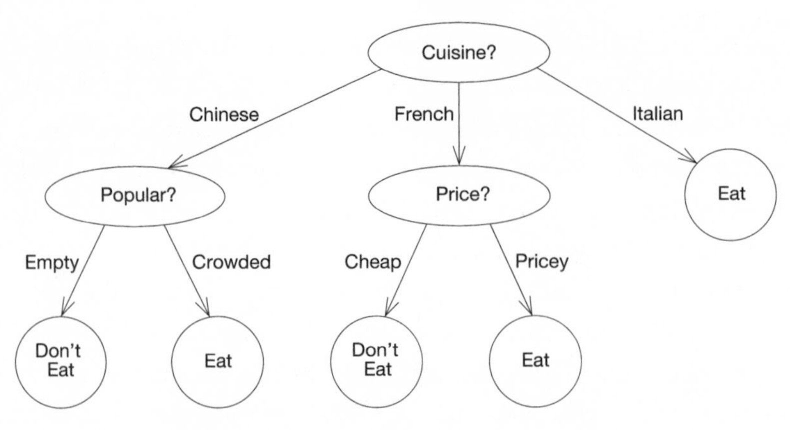


*A decision tree for choosing a restaurant*

learning in the first decade of the twenty-first century, and people used it for calculating everything from the subjects of news articles to the structures of proteins. Probabilistic models try to compute the likelihood of different possible answers and output the answer they think most probable, an approach that was vital for the success of IBM's Watson and that is likely to have continued impact.

Yet another approach, sometimes called genetic algorithms, is modeled on the process of evolution; different algorithms are tried out, and "mutated" in some form. The fittest* survive and propagate. Genetic algorithms have been used in applications ranging from designing radio antennas to playing video games, where they have sometimes achieved the same level of success as deep learning. And the list goes on; we won't go into all of them, preferring to focus on deep learning because it has become so dominant in recent years, but if you want to learn more about these various kinds of algorithms, Pedro Domingos's book *The Master Algorithm* is an excellent place to start. (With Domingos, we share the view that each has something to contribute, and also the view that the collection of algorithms has not yet been well integrated;

* Of course, what counts as fittest depends on what a particular system designer is trying to achieve; e.g., if the goal were mastery of a video game, the score on the game could be the measure of fitness.

later in the book we will say why we are less optimistic about finding a single master algorithm.) And many problems, like planning driving routes and robotic actions, still use techniques drawn from classical AI that use little or no machine learning. (Often, a single application, like the traffic routing algorithms used by Waze, will involve multiple techniques, drawn from both classical AI and machine learning.)

In recent years, machine learning has become ubiquitous, fueled in large part by big data. IBM's Watson relied on massive databases and used a mix of classical AI techniques and probabilistic machine learning to tune its system to win at *Jeopardy!* Cities use machine learning to allocate services, car-sharing services use machine learning to predict driver demand, and police departments use machine learning to predict crime. On the commercial side, Facebook uses machine learning to decide what news stories you might like to see in your feed, and also to infer what advertisements you might be likely to click on. Google uses machine learning to recommend videos, to place ads, to understand your speech, and to try to guess what you might be looking for with your web searches. Amazon's website uses machine learning to recommend products and interpret searches. Amazon's Alexa device uses machine learning to decode your requests, and so on.

None of these products is perfect; later we will discuss some examples of well-known commercial search engines flummoxed by basic requests. But virtually all of them are much better than nothing, and hence clearly economically valuable. Nobody could handwrite a search engine at the scale of the whole web; Google simply wouldn't exist without machine learning. Amazon product recommendations would be vastly inferior if they relied entirely on humans. (The closest thing we can think of is Pandora, a music recommendation service that is largely done by hand by human experts, but in consequence it is limited to a much smaller music library than comparable systems like Google Play that rely more heavily on machines.) Automated ad-recommendation systems that can personalize their recommendations to individual users based on historical statistics of what people with similar histories bought don't have to be perfect; even if they make the occasional mistake they are much more precisely targeted than the old-school strategy of simply placing big ads in a newspaper. In 2017, Google and Facebook, combined, made more than $80 billion placing ads; machine learning, based on statistical inference, was a central engine in making that happen.

o———o

Deep learning is based on two fundamental ideas.

The first, which might be called *hierarchical pattern recognition,* stems in part from a set of experiments in the 1950s that led neuroscientists David Hubel and Torsten Wiesel to win the 1981 Nobel Prize in Physiology or Medicine. Hubel and Wiesel discovered that different neurons in the visual system responded to visual stimuli in sharply different ways. Some responded most actively to very simple stimuli, like lines in particular orientations, while others responded more vigorously to more complex stimuli. The theory that they proposed was that complex stimuli might be recognized through a hierarchy of increasing abstraction, for instance, from lines to letters to words. In the 1980s, in a major landmark in AI history, Japanese neural network pioneer Kunahiko Fukushima built an actual computational realization of Hubel and Wiesel's idea, called Neocognitron, and showed that it could work for some aspects of computer vision. (Later books by Jeff Hawkins and Ray Kurzweil championed the same idea.)

Neocognitron was made up of a set of *layers* (which look like rectangles). Going from left to right in the diagram opposite, there is first an input layer that would be presented with a stimulus, essentially the pixels in a digital image; then subsequent layers that would analyze the image, looking for variations in contrast, edges, and so forth, ultimately culminating in an output layer that identifies what category the input belongs to. Connections between layers allowed all the relevant processing to take place. All of these ideas—

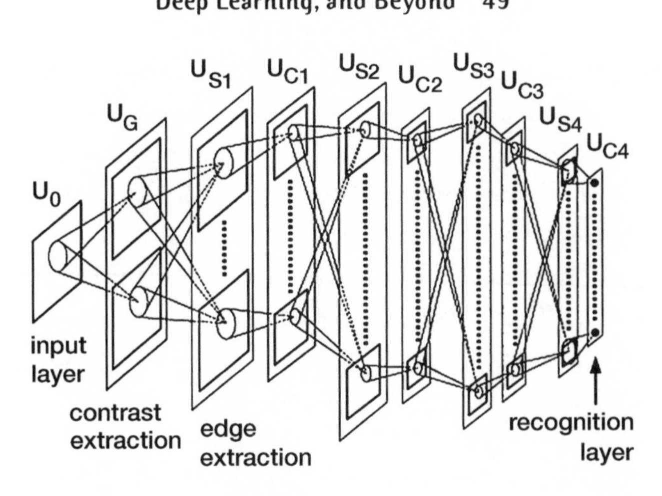


*Neocognitron, a neural network for recognizing objects*

input layers, output layers, and internal layers, with connections in between them—are now the backbone of deep learning.

Systems like these are called *neural networks,* because each layer consists of elements called *nodes* that can be (very loosely) likened to greatly simplified neurons. There are connections between those nodes, also called *connection weights,* or just weights; the larger the weight on the connection from node A to node B, the greater the influence of A on B. What a network does is a function of those weights.

o———o

The second key idea is learning. By strengthening the weights for a particular configuration of inputs to a particular output, for example, you could "train" a network to learn to associate a particular input to its corresponding output. Suppose, for example, that you want to get the network to learn the names of various letters presented on a grid of pixels. Initially the system would have no idea which pattern of pixels corresponded to which letter. Over time, through a process of trial and error and adjustment, it would come to associate pixels at the top of the grid with letters like T and E, and pixels on the left edge with letters like E, F, and H, gradually mastering the correlations between pixels in various locations and the appropriate labels. Back in the 1950s, Rosenblatt already had the core intuition that this could be viable, but the networks he used were simple and limited, containing only an input layer and an output layer. If your task was simple enough (like categorizing circles versus squares), some fairly straightforward mathematics guaranteed that you could always adjust your weights in order to "converge" on (compute) the right answer. But for more complicated tasks, two layers weren't enough to keep everything straight—intermediate layers representing combinations of things would be required—and at the time, nobody had a workable solution for reliably training deeper networks that had more than two layers. The primitive neural networks of the day had inputs (e.g., pictures) and outputs (labels) but nothing in between.

In their influential 1969 book *Perceptrons,* Marvin Minsky and Seymour Papert proved mathematically that simple two-layer networks weren't able to capture many things that systems might want to categorize.* Adding more layers, they observed, would give you more power, but would also come with a cost: the ability to train a network to find a satisfactory solution was no longer guaranteed. Somewhat pessimistically they wrote that their intuitive judgment was that the extension to multiple layers would be "sterile," though they left open the possibility that "perhaps some interesting convergence theorem [could] be discovered." Between that note of pessimism and the lack of truly compelling results, the field of neural networks soon withered. Simple problems were boring and of limited use, and complex problems seemed to be intractable.

But not everybody gave up. As Minsky and Papert acknowledged, they hadn't actually *proved* that you couldn't do anything useful at

* From a mathematical viewpoint, a two-layer network can identify features that fall on one side of a plane that divides a space of all possible inputs. Minsky and Papert proved that important basic geometric characteristics of an image, such as whether the image shows one object or two separate objects, couldn't be captured in this way.

all with a deeper network, only that you couldn't guarantee optimal results using the particular mathematics they reviewed. At one level, what Minsky and Papert wrote in 1969 still remains true: to this day, deep learning still doesn't offer much in the way of formal guarantees (except in the unrealistic case in which both infinite data and infinite computational resources are available); but in hindsight it is clear that Minsky and Papert underestimated how useful deeper networks could be without guarantees. Over the subsequent two decades, several people, including Geoff Hinton and David Rumelhart, independently invented mathematics that allows deeper neural networks to do a surprisingly good job, despite the lack of any formal mathematical guarantees of perfection.

o———o

A metaphor that people often use is of climbing up a hill. Imagine that the base of the mountain is a poor solution to a problem, in which the system has low accuracy, and the peak of the mountain is the best solution, with high accuracy. (The opposite metaphor, called *gradient descent,* is also sometimes used.)

What Hinton and others found was that although perfection could not be guaranteed in "deeper" networks with more than two layers, it was possible to build a system that could produce results that were often good enough, opportunistically climbing up the mountain by taking small steps of the right sort, using a technique called *backpropagation*—now the workhorse of deep learning.*

* Informally, the key idea in backpropagation revolves around finding a way to do "blame assignment" in complex networks. Suppose you are trying to train a neural network, by giving it a series of examples, say pictures with labels. Initially, your results will be poor, because (in the standard approach, anyway) all the weights are initially random; over time you want the weights to get tuned to numbers that fit your problem. In a two-layer network, it's obvious what to do: try out training examples, and see which weights are and are not helping you to get correct answers. If some of the connections between the nodes in the network's layers seem to contribute to the right answer, you make them stronger; if they contribute to a wrong

Backpropagation works by estimating the best direction up the mountain at any given point. Although finding its way to the highest summit of the mountain is not guaranteed—it can get stuck at "a local maximum" (a secondary peak, or even a large boulder on the mountain that is higher than anything nearby, but still not good enough)—in practice the technique is often adequate.

A technique called *convolution,* introduced in the late 1980s by Yann LeCun and still widely adopted, allowed object-recognition systems to be more efficient by building in an array of connections that would allow a system to recognize an object, no matter where it appears in the picture.

Still, although the math seemed sound, the initial results for neural networks weren't compelling. It was known that in principle, if you could find the right set of weights (a big but often manageable if), three or more layers could allow you to solve any problem you wanted, provided you had enough data and enough patience and an enormous number of nodes. But in practice it wasn't really enough; impossibly large numbers of nodes were needed, and the computers of the day couldn't make all the calculations you would need in reasonable amounts of time.

People had a strong hunch that more layers—which is to say "deeper networks"—would help. But nobody knew for sure. As late as the early 2000s, the hardware just wasn't up to the job. It would have taken weeks or even months of computer time to train a typi-

---

answer you make them weaker. In a two-layer network it is easy to identify which weights are contributing in which answers. Once you have a deeper network, with one or more "hidden layers" (called that because they are not connected directly to either the input or the output), it's less obvious which connections deserve credit and blame. Backpropagation came to the rescue.

Backpropagation works by calculating the difference between the desired output of a network and its actual output (known as the *error*) and then sending information about that error backward through the layers of the network, adjusting weights along the way, to improve performance on subsequent tests. With that bit of math, it became possible to train neural networks with three layers in a relatively reliable way.

cal deep network; you couldn't (as people now can) try a hundred different alternatives and sort through them to find the best one. Results were promising but not competitive with other approaches.

This is where GPUs came in. What finally catalyzed the deep learning revolution (aside from some important technical tweaks*) was figuring out how to harness GPUs efficiently in the service of training more complex models with more layers in reasonable time. Deep learning—which involves training networks with four or more layers, sometimes more than a hundred—was at last practical.

---

The results from deep learning have been truly remarkable. Researchers used to spend years crafting clever rules by hand trying to make object recognition work. Now that work can be replaced by a deep learning system that spends some hours or days computing. Deep learning has allowed people to take on new problems, too, not just ad recommendation, but also speech recognition and object recognition, that were never solved adequately using older machine-learning techniques.

Deep learning has achieved "state of the art" (best results to date) in benchmark after benchmark. For example, as a lengthy article in *The New York Times Magazine* explained, deep learning has radically improved Google Translate. Until 2016, Google Translate used classical machine-learning techniques, using enormous tables of matching patterns in the two languages, labeled with probabilities. A newer neural-network-based approach, leveraging deep learning,

* One technical tweak, for those who are interested, is called *dropout*, invented by Hinton and others, involved a way of dealing with overfitting, a problem in which a machine-learning algorithm learns the particular examples in the training data, but misses the general pattern that underlies them, like a student studying multiplication who just memorizes all the examples in his textbook but has no idea how to solve any new problems. Dropout tries to force the system to generalize, rather than merely memorize. Another tweak accelerated the part of the computation that related the output of a network node to its inputs.

yielded markedly better translations. Deep learning has also led to major improvements in getting machines to transcribe speech and label photographs.

Beyond that, in many (though not all) ways, deep learning is easier to use. Traditional machine learning often relies heavily on expertise in *feature engineering*. In vision, for example, skilled engineers who are knowledgeable about vision have tried to find common properties in visual images that might be helpful to machines trying to learn about images, such as edges, corners, and blobs. In 2011, what made someone a good machine-learning engineer was often the ability to find the right input features for a given problem.*

To some degree, deep learning has changed that; in many problems (though, as we will see, not all), deep learning can work well without extensive feature engineering. The systems that started winning the ImageNet competition learned to categorize objects—at state-of-the-art levels—without significant amounts of feature engineering. Instead the systems learned all they needed to know by just looking at the pixels in images and the labels they were supposed to learn. Feature engineering no longer seemed necessary. You didn't need a PhD in vision science to get started.

What's more, deep learning turned out to be astonishingly general, useful not only for problems in object recognition and speech recognition, but also for many tasks that people might have never even imagined were achievable. Deep learning has had remarkable success at creating synthetic art in the style of past masters (turn your landscape into a Van Gogh), and colorizing old pictures. It can be used, particularly in combination with an idea known as *generative adversarial networks,* to address the problem of "unsupervised learning," challenges in which there is no teacher to label examples.

* Sophisticated readers with expertise in the area will realize that rumors of the replacement of feature engineering have been somewhat exaggerated; the hard work that goes into crafting representations like Word2Vec still counts as feature engineering, just of a different sort than one has traditionally found in a field like computational linguistics.

Deep learning can also be used as a component in systems that play games, sometimes at superhuman levels. Both of DeepMind's initial signature successes—first on Atari video games, later on Go—relied in part on leveraging deep learning combined with *reinforcement learning* to yield a new technique known as *deep reinforcement learning,* a way of doing trial-and-error learning with large-scale data. (AlphaGo also borrowed some other techniques, as we will discuss later.)

At times the success can seem intoxicating. When Andrew Ng, a leading AI researcher who was then directing AI research at the Chinese search engine company Baidu, wrote in the pages of *Harvard Business Review* in 2016, "If a typical person can do a mental task with less than one second of thought, we can probably automate it using AI either now or in the near future," he was referring largely to the success of deep learning. The sky seemed like the limit.

Still, we have been skeptical from the start. Even if it was immediately evident that deep learning was a more powerful tool than any of its predecessors, it seemed to us that it was being oversold. In 2012, Gary wrote this in *The New Yorker,* driven by his research a dozen years earlier on predecessors to deep learning:

> Realistically, deep learning is only part of the larger challenge of building intelligent machines. Such techniques lack ways of representing causal relationships (such as between diseases and their symptoms), and are likely to face challenges in acquiring abstract ideas like "sibling" or "identical to." They have no obvious ways of performing logical inferences, and they are also still a long way from integrating abstract knowledge . . .

Several years later, all of this still appears to be true—despite all the manifest progress in certain areas, such as speech recognition, language translation, and voice recognition. Deep learning still isn't any kind of universal solvent. Nor does it have much to do with general intelligence, of the sort we need for open systems.

In particular, it faces three core problems, each of which affects both deep learning itself and other popular techniques such as deep reinforcement learning that rely heavily on it:

**Deep learning is greedy.** In order to set all the connections in a neural net correctly, deep learning often requires a massive amount of data. AlphaGo required 30 million games to reach superhuman performance, far more games than any one human would ever play in a lifetime. With smaller amounts of data, deep learning often performs poorly. Its forte is working with millions or billions of data points, gradually landing on a set of neural network weights that will capture the relations between those examples. If a deep learning system is given only a handful of examples, the results are rarely robust. And of course much of what we do as humans we learn in a just a few moments; the first time you are handed a pair of 3-D glasses you can probably put them on and infer roughly what is going on, without having to try them on a hundred thousand times. Deep learning simply isn't built for this kind of rapid learning.

Where Andrew Ng promised that machines might soon automate anything a person can do in a second, a more realistic view might be "If a typical person can do a mental task with less than one second of thought, and we can gather an enormous amount of directly relevant data, we have a fighting chance—so long as the problems we actually encounter aren't too terribly different from the training data, and the domain doesn't change too much over time."

That's fine for a board game like Go or chess with rules that have been static for millennia, but as we noted in the introduction, in many real-world problems getting enough data of the right sort is unrealistic or even impossible. A big part of why deep learning struggles in language, for example, is that in language, new sentences with fresh meanings are in infinite supply, each one subtly different from the last. The more different your real-world problems are from the data you used to train the system, the less likely the system is to be reliable.

**Deep learning is opaque.** Whereas classical expert systems are made up of rules that can be relatively easily understood (like "If a person has an elevated white blood cell count, then it is likely that the person has an infection"), neural networks are made up of vast arrays of numbers, virtually none of which make intuitive sense to ordinary human beings. Even with sophisticated tools, experts struggle to understand why particular neural networks make the decisions they do. There is an unsolved mystery about why neural networks work as well as they do, and a lack of clarity about the exact circumstances in which they don't. A neural network that learns a particular task might score, say, 95 percent correct on some test. But then what? It's often very difficult to figure out *why* the network is making mistakes on the other 5 percent, even when some of those mistakes are dramatic errors that no human would ever make, like the confusion we saw earlier between a refrigerator and a parking sign. And if those mistakes matter and we can't understand why the system is making them, we have a problem.

The problem is particularly acute since neural networks can't give human-style explanations for their answers, correct or otherwise.* Instead neural networks are "black boxes"; they do what they do, and it is hard to understand what's inside. To the extent that we might wish to rely on them for things like driving or performing household tasks, that's a serious problem. It's also a problem if we want to use deep learning in the context of larger systems, because we can't really tell the operating parameters—engineer's lingo for when things work and don't. Prescription medicines come with all kinds of information about which potential side-effects are dangerous and which are merely unpleasant, but someone selling a deep-learning-based face-recognition system to a police department may not be able to tell the customer much in advance about when it will work and when it won't. Maybe it will turn out that it works reasonably well with Caucasians on sunny days, but fails

* Of course if the system were perfect, and we could count on it, we wouldn't need to look inside, but few current systems are perfect.

with African Americans in dim light; it is difficult to know, except through experimentation.

A further consequence of the opacity of deep learning is that it doesn't fit naturally with knowledge about how the world works. There is no easy way to tell a deep network "apples grow on trees" or "apples tend to fall down, not up, when they fall out of trees." This is fine if all you have to do is recognize an apple; it is devastating if you want a deep learning system to interpret what happens in a Rube Goldberg contraption when a ball falls off a ramp, down a chute, and onto an escalator.

**Deep learning is brittle.** As we saw in the opening chapter, deep learning can be perfect in one situation and utterly wrong in another. Hallucinations like the nonexistent refrigerator we saw earlier aren't some sort of one-off that we are picking on; they remain a persistent problem, years after the problem was first noticed. More recent models continue to be subject to the same thing, making such errors 7 to 17 percent of the time, according to one study. In a typical example, a photo shows two smiling women chatting on cell phones with some out-of-focus trees behind them, with one woman facing the camera, the other turned so only her cheek was visible. The resulting caption—*A woman talking on a cell phone while sitting on a bench*—is a mixture of partly accurate detail and total confabulation, perhaps driven by statistical quirks in the training set; one of the women had disappeared, and the bench came from nowhere. It is not hard to see how similar mechanisms could lead an automated security robot to misconstrue a cell phone as a gun.

There are literally dozens of ways to fool deep

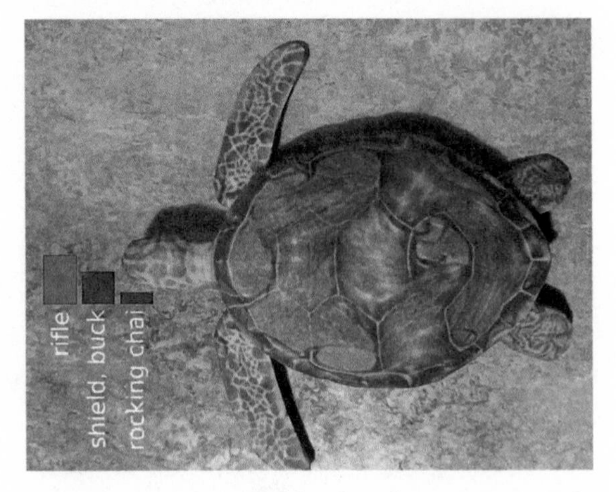


*A turtle, misidentified by deep learning as a rifle*

*A baseball with foam, misidentified as an espresso*

networks. Some MIT researchers, for example, designed a three-dimensional turtle that a deep learning system mistook for a rifle.

And sticking the turtle underwater (where you usually wouldn't find a rifle) didn't help. Likewise, the team put a little foam on a baseball nestled in a brown glove, and the network thought the ball was an espresso—from every angle—even when the ball was right in front of a baseball glove.

Another team of researchers added little patches of random noise unobtrusively in tiny corners of images, and tricked neural nets into thinking that a piggy bank was a "tiger cat."

*A psychedelic toaster*

Another team added small stickers with psychedelic pictures of toasters to real-world objects like a banana and fooled a deep learning system into thinking that this scene was a toaster, rather than, say, a banana next to a tiny, distorted picture of a toaster. If your son or daughter missed the banana, you would rush them to a neurologist.

*A banana, correctly identified by deep learning*

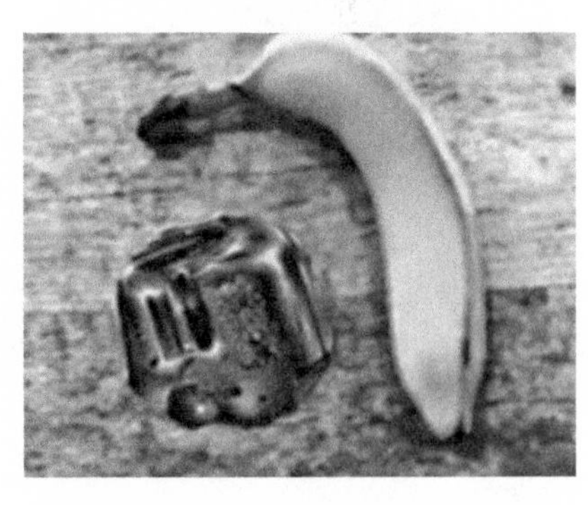

*A banana with sticker, misidentified as a toaster*

And then there is this deliberately altered stop sign, which a deep learning system misread as a speed limit sign.

*An altered stop sign, misidentified as a speed limit sign*

Still another research team compared deep learning systems to humans on twelve different tasks in which images were distorted in a dozen different ways, such as turning color images to black and white, replacing colors, rotating the image, and so forth. Humans almost invariably performed better. Our visual systems are built to be robust; deep learning, not so much.

Another study showed that deep learning has trouble recognizing ordinary objects in unusual poses, like this flipped-over schoolbus that was mistaken for a snowplow.

*A schoolbus in an unusual position, misidentified as a snowplow*

Things get even quirkier when we get to language. Stanford computer scientists Robin Jia and Percy Liang did a study of systems that work on the SQuAD task, mentioned in chapter 1, in which deep learning systems tried to underline the answers to questions in text. Deep learning systems were given this text—

> Peyton Manning became the first quarterback ever to lead two different teams to multiple Super Bowls. He is also the oldest quarterback ever to play in a Super Bowl at age 39. The past record was held by John Elway, who led the Broncos to victory in Super Bowl XXXIII at age 38 and is currently Denver's Executive Vice President of Football Operations and General Manager.

and this question—

> What is the name of the quarterback who was 38 in Super Bowl XXXIII?

One deep learning system correctly underlined "John Elway." So far, so good. However, when Jia and Liang gave the system the same paragraph but added at the end one additional sentence, of totally irrelevant information—"Quarterback Jeff Dean had jersey number 17 in Champ Bowl XXXIV"—and then asked the same question about the Super Bowl, the system was totally confused, attributing the win to Jeff Dean rather than John Elway, conflating the two sentences about two different championships, and showing no genuine understanding of either.

Another study showed how easy it is to fake out question-answering systems by asking partial questions. Deep learning systems' reliance on correlation rather than understanding regularly leads them to hit the buzzer with a random guess even if the question hasn't been finished. For instance, if you ask the system "How many," you get the answer "2"; if you ask "What sport?" you get the answer "tennis." Play with these systems for a few minutes, and you have the sense of interacting with an elaborate parlor trick, not a genuine intelligence.

An even weirder version of the same problem can occur in machine translation. When Google Translate was given the input text *dog dog dog dog dog dog dog dog dog dog dog dog dog dog dog dog dog dog* and asked to translate that from Yoruba (and some other languages) to English, it gave back this translation:

> Doomsday clock is three minutes to twelve. We are experiencing characters and a dramatic developments in the world, which indicate that we are increasingly approaching the end times and Jesus' return.

In the end, deep learning just ain't that deep. It is important to recognize that, in the term *deep learning,* the word "deep" refers to the number of layers in a neural network and nothing more. "Deep," in that context, doesn't mean that the system has learned anything particularly conceptually rich about the data that it has seen. The deep reinforcement learning algorithm that drives DeepMind's Atari game system, for example, can play millions of games of *Breakout,* and still never really learn what a paddle is, as was elegantly demonstrated recently by the AI startup Vicarious. In *Breakout,* the player moves the paddle back and forth along a horizontal line. If you change the game so that the paddle is now a few pixels closer to the bricks (which wouldn't bother a human at all), DeepMind's entire system falls apart. (A team at Berkeley showed something similar with *Space Invaders:* tiny bits of noise shattered performance, showing just how superficial the system for learning games really was.)

Some people within the field seem finally to be recognizing these points. University of Montreal professor Yoshua Bengio, one of the pioneers of deep learning, recently acknowledged that "deep [neural networks] tend to learn surface statistical regularities in the dataset rather than higher-level abstract concepts." In an interview near the end of 2018, Geoff Hinton and Demis Hassabis, founder of DeepMind, similarly suggested that general artificial intelligence was nowhere close to being a reality.

One winds up with a kind of "long tail" problem, in which there are a few common cases (the fat head of a collection of data) that are easy, but also a large number of rare items (the long tail) that are very hard to learn. It's easy to get deep learning to tell you that a group of young kids are playing Frisbee, because there are so many labeled examples. It's much harder to get it to tell you what's out of the ordinary about this picture:

*Familiar objects, in an unusual pose*

Dogs, cats, toy horses, and carriages are all perfectly common, but this particular configuration of elements isn't, and deep learning has no idea what to do.

o———o

So why has deep learning been so oversold, given all these problems? It's obviously effective for statistical approximation with large data sets, and it has a certain kind of elegance—one simple equation that seems to solve so much. It also has considerable commercial value. In hindsight, however, it's just as obvious that something is missing.

We see all the exuberance for deep learning as an example of the illusory progress gap that we discussed in the opening chapter; on some tasks, deep learning can succeed, but that doesn't mean that a real intelligence lies behind the behavior.

Deep learning is a very different beast from a human mind. At best, deep learning is a kind of idiot savant, with miraculous perceptual abilities, but very little overall comprehension. It's easy to find effective deep learning systems for labeling pictures—Google, Microsoft, Amazon, and IBM, among others, all offer commercial systems that do this, and Google's neural network software library TensorFlow allows any computer science student the chance to do it for free. It's also easy to find effective deep learning systems for speech recognition, again more or less a commodity at this point. But speech recognition and object recognition aren't intelligence, they are just slices of intelligence. For real intelligence you also need reasoning, language, and analogy, none of which is nearly so well handled by current technology. We don't yet, for example, have AI systems that can reliably understand legal contracts, because pattern classification on its own isn't enough. To understand a legal contract you need to be able to reason about what is and isn't said, how various clauses relate to previously established law, and so forth; deep learning doesn't do any of that. Even reliably summarizing plots of old movies for Netflix would be too much to ask.

Indeed, even in the slice of cognition that is perception, which comes closest to being deep learning's forte, progress is only partial; deep learning can identify objects, but it can't understand the relations between them, and often it can be fooled. In other areas, like language understanding and everyday reasoning, deep learning lags far behind human beings.

A vast majority of what has been written about deep learning in the popular media makes it seem as if progress in one of these areas is tantamount to progress in all of them. *MIT Technology Review,* for example, put Deep Learning on its 2013 annual list of breakthrough technologies, and summarized it like this:

> With massive amounts of computational power, machines can now recognize objects and translate speech in real time. Artificial intelligence is finally getting smart.

But the logic is flawed; just because you recognize a syllable or a border collie doesn't mean you're smart. Not all cognitive problems are created equal. To extrapolate from success in one aspect of cognition to success in all aspects of cognition is to succumb to the illusion of progress.

In the final analysis, deep learning is both beautiful and tragic. It is beautiful because on a good day it requires very little work; you often don't need to spend much time on the pesky job of feature engineering, and in the best case the machine takes care of a large fraction of what needs to be done. It is tragic because nothing ever guarantees that any system in the real world will give you the right answer when you need it, or ensures that you will be able to debug it if it doesn't work. Instead it's in some ways more like an art rather than science; you try things out and if you have enough data they tend to work. But you can't prove this in advance, the way you might prove some theorem about geometry. And no theory predicts precisely which tasks deep learning can and can't robustly solve. You often have to try it out, empirically. You see what works and what doesn't, often tinkering with your initial system and your data set, until you get the results you want. Sometimes that's easy. Sometimes it's hard.

Deep learning is an extremely valuable tool for AI—we expect that it will continue to play an important and integral role going forward, and that people will invent many other creative applications for it that we haven't even thought of. But it's much more likely to be just one component in an array of tools than a stand-alone solution.

Ultimately what has happened is that people have gotten enormously excited about a particular set of algorithms that are terrifically useful, but that remain a very long way from genuine

This *is your machine learning system?* (Randall Munroe, xkcd.com)

intelligence—as if the discovery of a power screwdriver suddenly made interstellar travel possible. Nothing could be further from the truth. We need the screwdriver, but we are going to need a lot more, too.

By this we don't mean that deep learning systems can't do things that appear intelligent, but we do mean that deep learning on its own lacks the flexibility and adaptive nature of real intelligence. In the immortal words of Law 31 of Akin's Laws of Spacecraft Design, "You can't get to the moon by climbing successively taller trees."

In the rest of the book, we will describe what it will take to get to the moon—or at least get to machines that can think and reason and talk and read with the versatility of your average human being. What we need is not just "deeper" learning, in the sense of having more layers in a neural network, but *deeper understanding*. We need systems that can truly reason about the complex interplay of entities that causally relate to one another in an ever-changing world.

To see what we mean by that, it's time to dig deep, into two of the most challenging domains in AI: reading and robots.

4

# If Computers Are So Smart, How Come They Can't Read?

SAMANTHA: So how can I help you?
THEODORE: Oh, it's just more that everything feels disorganized, that's all.
SAMANTHA: You mind if I look through your hard drive?
THEODORE: Um . . . okay.
SAMANTHA: Okay, let's start with your e-mails. You have several thousand e-mails regarding *LA Weekly,* but it looks like you haven't worked there in many years.
THEODORE: Oh, yeah. I think I was just saving those cause, well I thought maybe I wrote something funny in some of them. But . . .
SAMANTHA: Yeah, there are some funny ones. I'd say that there are about eighty-six that we should save, we can delete the rest.

—*HER* (2013), WRITTEN AND DIRECTED BY SPIKE JONZE

Wouldn't it be nice if machines could understand us as well as Samantha (the "operating system" voiced by Scarlett Johansson in the science fiction movie *Her*) understands Theodore? And if they could sort through our emails in an instant, pick out whatever we need, and filter out the rest?

If we could give computers one gift that they don't already have, it would be the gift of understanding language, not just so they could help organize our lives, but also so they could help humanity on some of our greatest challenges, like distilling vast scientific literatures, that individual humans can't possibly hope to keep up on.

In medicine, seven thousand papers are published every day. No doctor or researcher can possibly read them all, and that is a serious impediment to progress. Drug discovery gets delayed in part because lots of information is locked up in literature that nobody has time to read. New treatments sometimes don't get applied because doctors don't have time to read and discover them. AI programs that could automatically synthesize the vast medical literature would be a true revolution.

Computers that could read as well as PhD students but with the raw computational horsepower of Google would revolutionize science, too. We would expect advances in every field, from mathematics to climate science to material science. And it's not just science that would be transformed. Historians and biographers could instantly find out everything that has been written about an obscure person, place, or event. Writers could automatically check for plot inconsistencies, logical gaps, and anachronisms.

Even much simpler abilities could be enormously helpful. Currently the iPhone has a feature such that when you get an email message that sets up an appointment, you can click on it and the iPhone will add it to your calendar. That's really handy—when it works right. Often, it doesn't; the iPhone adds the appointment, not on the day you have in mind, but perhaps on some other day mentioned in the email. If you don't catch the mistake when the iPhone makes it, it can be a disaster.

Someday, when machines really can read, our descendants will wonder how we ever got by without synthetic readers, just as we wonder how earlier generations managed without electricity.

o———o

At TED, in early 2018, the noted futurist and inventor Ray Kurzweil, now working at Google, announced his latest project, Google Talk to Books, which promised to use natural language understanding to "provide an entirely new way to explore books." *Quartz* dutifully hyped it as "Google's astounding new search tool [that] will answer any question by reading thousands of books."

As usual, the first question to ask is "What does the program actually do?" The answer was that Google has indexed the sentences in 100,000 books, ranging from *Thriving at College* to *Beginning Programming for Dummies* to *The Gospel According to Tolkien,* and developed an efficient method for encoding the meanings of sentences as sets of numbers known as vectors. When you ask a question, it uses those vectors to find the twenty sentences in the database that have vectors that are most similar. The system has no idea what you are actually asking.

Just from knowing the input to the system it should be immediately obvious that the claim in the *Quartz* article that Talk to Books "will answer any question" can't be taken literally; 100,000 books may sound like a large number but it's a tiny fraction of the more than one hundred million that have been published. Given what we saw earlier about how much deep learning draws on correlation rather than genuine comprehension, it should come as no surprise that many of the answers were dubious. If you asked it some particular detail of a novel, for instance, you should reasonably expect a reliable answer. Yet when we asked "Where did Harry Potter meet Hermione Granger?" none of the twenty answers was from *Harry Potter and the Sorcerer's Stone,* and none addressed the question itself: where the meeting took place. When we asked "Were the Allies justified in continuing to blockade Germany after World War I?" it found no results that even mentioned the blockade. Answering "any question" is a wild exaggeration.

And when answers weren't spelled out directly in a phrase in the indexed text, things often ran amiss. When we asked "What were the seven Horcruxes in *Harry Potter*?" we didn't even get an answer with a list, perhaps because none of the many books that discuss Harry Potter enumerates the Horcruxes in a single list. When we asked "Who was the oldest Supreme Court justice in 1980?" the system failed, even though you as a human can go to any online list of the Supreme Court justices (for instance, in Wikipedia) and in a couple of minutes figure out that it was William Brennan. Talk to Books stumbled again, precisely because there was no sentence spelling it out in full—"The oldest judge on the Supreme Court in 1980 was William Brennan"—in any of the 100,000 books, and it had no basis for making inferences that extended beyond the literal text.

The most telling problem, though, was that we got totally different answers depending on how we asked the question. If we asked Talk to Books, "Who betrayed his teacher for 30 pieces of silver?," a pretty famous incident in a pretty famous story, of the twenty answers, only six correctly identified Judas. (Curiously, nine of the answers had to do with the much more obscure story of Micah the Ephraimite, in Judges 17.) But things got even worse as we strayed from the exact wording of "pieces of silver." When we asked Talk to Books the slightly less specific "Who betrayed his teacher for 30 coins?" Judas turned up in only 10 percent of the answers. (The top-ranked answer was both irrelevant and uninformative: "It is not known who Jingwan's teacher was.") And when we again slightly reworded the question, this time changing "betrayed" to "sold out," yielding "Who sold out his teacher for 30 coins?" Judas disappeared from the top twenty results altogether.

The further we moved from exactly matching a set of words, the more lost the system became.

---

The machine-reading systems of our dreams, when they arrive, would be able to answer essentially any reasonable question about what they've read. They would be able to put together information across multiple documents. And their answers wouldn't just consist of spitting back underlined passages, but of *synthesizing* information, whether that's lists of Horcruxes that never appeared in the same passage, or the sort of pithy encapsulations that you would expect of a lawyer assembling precedents across multiple cases, or a scientist formulating a theory that explains observations collected across multiple papers. Even a first grader can create a list of all the good guys and bad guys that appear in a series of children's books.

Just as a college student writing a term paper can bring together ideas from multiple sources, cross-validating them and reaching novel conclusions, so too should any machine that can read.

But before we can get machines to synthesize information rather than merely parroting it, we need something much simpler: machines that can reliably comprehend even basic texts.

That day isn't here yet, however excited some people seem to be about AI. To get a sense for why robust machine reading is actually still a fairly distant prospect, it helps to appreciate—in detail—what is required even to comprehend something relatively simple, like a children's story.

Suppose that you read the following passage from *Farmer Boy,* a children's book by Laura Ingalls Wilder (author of *Little House on the Prairie*). Almanzo, a nine-year-old boy, finds a wallet (then referred to as a "pocketbook") full of money dropped in the street. Almanzo's father guesses that the "pocketbook" (i.e., wallet) might belong to Mr. Thompson, and Almanzo finds Mr. Thompson at one of the stores in town.

> Almanzo turned to Mr. Thompson and asked, "Did you lose a pocketbook?"
>
> Mr. Thompson jumped. He slapped a hand to his pocket, and fairly shouted.
>
> "Yes, I have! Fifteen hundred dollars in it, too! What about it? What do you know about it?"
>
> "Is this it?" Almanzo asked.
>
> "Yes, yes, that's it!" Mr. Thompson said, snatching the pocketbook. He opened it and hurriedly counted the money. He counted all the bills over twice. . . .
>
> Then he breathed a long sigh of relief and said, "Well, this durn boy didn't steal any of it."

A good reading system should be able to answer questions like these:

- Why did Mr. Thompson slap his pocket with his hand?
- Before Almanzo spoke, did Mr. Thompson realize that he had lost his wallet?
- What is Almanzo referring to when he asks "Is this it?"
- Who almost lost $1,500?
- Was all of the money still in the wallet?

All of these questions are easy for people to answer, but no AI yet devised could reliably handle queries like these. (Think about how troubled Google Talk to Books would have been by them.)*

At its core, each of these questions requires a reader (human or otherwise) to follow a chain of inferences that are only implicit in the story. Take the first question. Before Almanzo speaks, Mr. Thompson doesn't know he has lost the wallet and assumes that he has the wallet in his pocket. When Almanzo asks him whether he has lost a wallet, Thompson realizes he might in fact have lost his wallet. It is to test this possibility—the wallet might be lost—that Thompson slaps his pocket. Since the wallet isn't where he usually keeps it, Thompson concludes that he has lost his wallet.

When it comes to complex chains of reasoning, current AI is at a loss. Such chains of reasoning often demand that the reader put together an impressive range of background knowledge about people and objects, and more generally about how the world works,

* The Allen Institute for Artificial Intelligence has a website, ai2.org, where you can try out near-state-of-the-art models on tests like these. For example, on November 16, 2018, we entered the Almanzo story into the then leading model available on the site and asked *How much money was in the pocketbook?*, *What was in the pocketbook?*, *Who owns the pocketbook?*, and *Who found the pocketbook?* The first and third were answered correctly; the second received an incoherent answer ("counted the money"); and the last was simply wrong (Mr. Thompson, rather than Almanzo). Unreliable results like these are typical of the contemporary state of the art.

and no current system has a broad enough fund of general knowledge to do this well.

Take some of the kinds of knowledge you probably drew on just now, automatically, without even being aware of it, as you digested the story of Almanzo and the wallet:

- People can drop things without realizing it. *This is an example of knowledge about the relation between events and people's mental states.*
- People often carry their wallets in their pockets. *This is an example of knowledge about how people typically use certain objects.*
- People often carry money in their wallets, and money is important to them, because it allows them to pay for things. *This is an example of knowledge about people, customs, and economics.*
- If people assume that something important to them is true, and they find out that it might not be true, then they often urgently try to verify it. *This is an example of knowledge about the kinds of things that are psychologically important to people.*
- You can often find out whether something is inside your pocket by feeling the outside of the pocket. *This is an example of how different types of knowledge may be combined. Here knowledge about how different objects (hands, pockets, wallets) interact with one another is combined with knowledge about how the senses work.*

The reasoning required for the other questions is equally rich. To answer the third question, "What is Almanzo referring to when he asks 'Is this it?,' " the reader has to understand something about language, as well as about people and objects, concluding that a reasonable antecedent to the words "this" and "it" could be the wallet, but (rather subtly) that "this" refers to the wallet that Almanzo is holding, while "it" refers to the wallet that Mr. Thompson has lost.

Happily, the two (what Almanzo is holding and what Mr. Thompson has lost) turn out to be the same.

To cope with even a simple passage, one's knowledge of people, objects, and language must be deep, broad, and flexible; if circumstances are even slightly different, we need to adapt accordingly. We should not expect equal urgency from Mr. Thompson if Almanzo said that he had found Almanzo's grandmother's wallet. We find it plausible that Mr. Thompson could have lost his wallet without knowing it, but we would be surprised if he was unaware of having his wallet taken after he was mugged at knifepoint. Nobody has yet been able to figure out how to get a machine to reason in such flexible ways. We don't think this is impossible, and later we sketch some of the steps that would need to be taken, but the reality for now is that what is required vastly outstrips what any of us in the AI community have yet managed to accomplish. Google Talk to Books wouldn't even be close (nor would the readers from Microsoft and Alibaba that we mentioned at the very beginning of the book).

Fundamentally, there is a mismatch between what machines are good at doing now—classifying things into categories—and the sort of reasoning and real-world understanding that would be required in order to capture this mundane yet critical ability.

---

Virtually anything you might read poses similar challenges. There's nothing particularly special about the Wilder passage. Here's a brief example from *The New York Times*, April 25, 2017.

> Today would have been Ella Fitzgerald's 100th birthday.
>
> One New Yorker, Loren Schoenberg, played saxophone alongside the "First Lady of Song" in 1990, near the very end of her career. He compared her to "a vintage bottle of wine" . . .

Anyone can easily answer questions taken pretty directly from the text (*what instrument did Loren Schoenberg play?*)—but many questions would demand a kind of inference that entirely eludes most current AI systems.

- Was Ella Fitzgerald alive in 1990?
- Was she alive in 1960?
- Was she alive in 1860?
- Did Loren Schoenberg ever meet Ella Fitzgerald?
- Does Schoenberg think that Fitzgerald was an alcoholic beverage?

*"He compared her to 'a vintage bottle of wine.'"*

Answering the first, second, and third questions involves reasoning that Ella was born on April 25, 1917, since April 25, 2017, was her 100th birthday, and then incorporating common knowledge such as facts that

- People are alive during their career, so she was alive in 1990.
- People are alive at all times between their birth and their death, and no times before their birth or after their death. So

> Fitzgerald must have been alive in 1960 and not yet alive in 1860.

Answering the fourth question involves reasoning that playing music alongside someone generally involves meeting them, and inferring that Fitzgerald is "the First Lady of Song," even though that identity is never quite made explicit.

Answering the fifth question requires reasoning about what sorts of things people typically envision when they make comparisons, and knowing that Ella Fitzgerald was a person and that people cannot turn into beverages.

Pick a random article in the newspaper, or a short story, or novel of any length, and you are sure to see something similar; skilled writers don't tell you everything, they tell you what you need to know, relying on shared knowledge to fill in the gaps. (Imagine how dull Wilder's story would be if she had to tell you that people keep their wallets in their pockets, and that people sometimes attempt to detect the presence or absence of small physical objects by reaching for them with their hands, through their pockets.)

In an earlier era, a bunch of AI researchers actually tried hard to solve these problems. Peter Norvig, now a director of research at Google, wrote a provocative doctoral thesis on the challenges in getting machines to understand stories. More famously, Roger Schank, then at Yale, came up with a series of insightful examples of how machines could use "scripts" to understand what happens when a customer goes to a restaurant. But understanding a story requires much more complex knowledge and many more forms of knowledge than scripts, and the problem of formulating and collecting all that knowledge was daunting. In time, the field gave up, and researchers started working on other, more approachable problems—such as web search and recommendation engines—none of which has brought us significantly closer to general AI.

Web search has of course nonetheless changed the world; it's one of AI's biggest success stories. Google Search, Bing, and others are amazingly powerful and fantastically useful pieces of engineering, powered by AI, that almost instantaneously find matches among billions of web documents.

What is perhaps surprising is that, while they are all powered by AI, they have almost nothing to do with the kind of automated, synthetic machine reading we have been calling for. We want machines that can understand what they are reading. Search engines don't.

Take Google Search. There are two basic ideas in the Google algorithm, one old, and one that Google pioneered. Neither depends on having the system comprehend documents. The first, older idea had been used in document-retrieval programs since the early 1960s, long before Google or the web: you match words in the query against words in the document. Want to search for recipes involving cardamom? No problem—just find all the websites containing the words "recipe" and "cardamom." No need to understand that cardamom is a spice, no need to understand what it smells like, or tastes like, nor to know anything about the history of how it is extracted from pods or which cuisines most often use it. Want to find instructions on building airplanes? Just match a few words like "model," "airplane," and "how to," and you will get lots of useful hits, even if the machine has no idea what an airplane actually is, let alone what lift and drag are or the reasons you would probably rather fly commercial than get a ride on a scale model.

The second, more innovative idea—the famous PageRank algorithm—was the idea that a program could use the collective wisdom of the web in judging which web pages were high quality by seeing which pages had gotten many links, particularly links from other high-quality pages. That insight catapulted Google above all the other web search engines of the time. But matching words does not have much to do with *understanding* texts, nor does counting links that are inbound from other pages.

The reason that Google Search does as well as it does *without* any kind of sophisticated reading is that little precision is required. The search engine does not need to read deeply in order to discern whether some treatise on presidential powers leans to the left or the right; the user can figure that out. All Google Search has to figure out is whether a given document is about the right general topic. One can usually get a pretty good idea of the subject of a document just by looking at the words and short phrases that are in it. If it has "president" and "executive privilege," the user probably will be happy to have the link; if it's about the Kardashians, it's probably not relevant. If a document mentions "George," "Martha," and the "Battle of Yorktown," Google Search can guess that the document is about George Washington, even if it knows nothing about marriage or revolutionary wars.

---

Google is not always so superficial. Sometimes it manages to interpret queries and give back fully formed answers rather than just long lists of links. That's a little closer to reading, but only a little, because Google is generally only reading the queries, not the documents themselves. If you ask "What is the capital of Mississippi?" Google correctly parses your question and looks up the answer ("Jackson") in a table that's been constructed in advance. If you ask "How much is 1.36 euros in rupees," parsing is again correct and the system can, after consulting a different table (this time with exchange rates), correctly calculate that "1.36 euros = 110.14 Indian rupees."

For the most part, when Google returns an answer of this sort, it's usually reliable (the system presumably only does so when its indicators suggest the answers are likely to be correct). But it's still far from perfect, and the errors it makes give a good hint about what's going on. For example, in April 2018, we asked Google Search "Who is currently on the Supreme Court?" and got back the rather incomplete answer "John Roberts," just one member among nine. As a bonus, Google provided a list of seven other justices "people also search for": Anthony Kennedy, Samuel Alito, Clarence Thomas,

Stephen Breyer, Ruth Bader Ginsburg, and Antonin Scalia. All these people have of course been on the court, but Scalia was deceased. Scalia's successor, Neil Gorsuch, and recent appointees Elena Kagan and Sonia Sotomayor were absent from Google's list. It's almost as if Google had missed the word "currently" altogether.

Going back to our earlier point about synthesis, the ultimate machine-reading system would compile its answer by reading Google News and updating its list when there are changes; or at least by consulting Wikipedia (which humans update fairly regularly) and extracting the current judges. Google doesn't seem to be doing that. Instead, as best we can tell, it is simply looking at statistical regularities (Alito and Scalia come up in many searches for justices), rather than genuinely reading and comprehending its sources.

To take another example, we tried asking Google, "When was the first bridge ever built?" and got back the following at the top of the results:

> Iron and Steel bridges are used today and most of the worlds [*sic*] major rivers are crossed by this type. The picture shows the first iron bridge in the world. It was built in Telford in 1779 by Abraham Darby (the third) and was the first large structure in history to be constructed from iron.

The words "first" and "bridge" match our query, but the first bridge ever built wasn't iron, and "first iron bridge" doesn't equal "first bridge"; Google was off by thousands of years. And the fact is, more than a decade after they were introduced, searches in which Google reads the question and gives a direct answer still remain very much in the minority. When you get links rather than answers, it's generally a sign that Google is just relying on things like keywords and link-counting, rather than genuine comprehension.

Companies like Google and Amazon are of course constantly improving their products, and it's easy enough to hand-code a system to correctly list the current set of Supreme Court justices; small incremental improvements will continue. What we don't see on the horizon is any *general* solution to the many kinds of challenges we have raised.

A few years ago, we saw a clever Facebook meme: a picture of Barack Obama with the caption "Last year you told us you were 50 years old; now you say you are 51 years old. Which is it, Barack Obama?" Two different utterances, spoken at different times, can both be true. If you're human, you get the joke. If you are a machine doing little more than keyword matching, you are lost.

---

What about speech-driven "virtual assistants" such as Siri, Cortana, Google Assistant, and Alexa? On the plus side, they often take action rather than merely giving you lists of links; unlike Google Search, they have been designed from the beginning to interpret user queries not as collections of random keywords, but as actual questions. But after several years, all are hit-or-miss, effective in some domains and weak in others. For example, they are all pretty good at "factoid" questions—"Who won the World Series in 1957?"; each of them also has pockets of clear strength. Google Assistant is good at giving directions and buying movie tickets. Siri is good at giving directions and at making reservations. Alexa is good at math, pretty decent at telling prewritten jokes, and (not surprisingly) good at ordering things from Amazon.

But outside their particular areas of strength, you never know what to expect. Not long ago, the writer Mona Bushnell tried asking all four programs for directions to the nearest airport. Google Assistant gave her a list of travel agents. Siri gave her directions to a seaplane base. Cortana gave her a list of airline ticket websites, such as Expedia. On a recent drive that one of us took, Alexa scored 100 percent on questions like *Is Donald Trump a person?*, *Is an Audi a vehicle?*, and *Is an Edsel a vehicle?*, but bombed on questions like *Can an Audi use gas?*, *Can an Audi drive from New York to California?*, and *Is a shark a vehicle?*

Or take this example, sent to Gary recently on Twitter: a screenshot of someone's effort to ask Siri for "the nearest fast food restaurant that was not McDonald's." Siri dutifully came up with a list of three nearby restaurants, and all served fast food—but every one of them was a McDonald's; the word "not" had been entirely neglected.

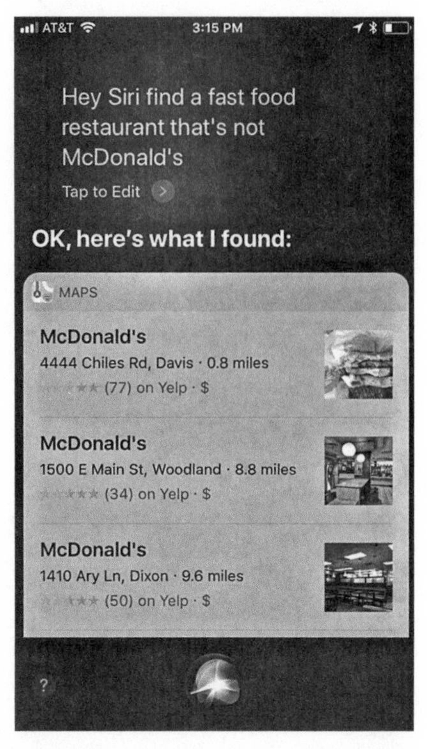


*Misunderstanding "Find a fast food restaurant that is not McDonald's"*

WolframAlpha, introduced in 2009 to much hype as "the world's first computational knowledge engine," is no better. It has enormous built-in databases of all kinds of scientific, technological, mathematical, census, and sociological information, and a collection of techniques for using this information to answer questions, but its capacity to put all that information together is still spotty.

Its strength is mathematical questions like "What is the weight of a cubic foot of gold?" "How far is Biloxi, Mississippi, from Kolkata?" and "What is the volume of an icosahedron with edge length of 2.3 meters?" ("547 kg," "8781 miles," and "26.5 m$^3$," respectively.)

But the limits of its understanding aren't hard to reach. If you ask "How far is the border of Mexico from San Diego?" you get "1144 miles," which is totally wrong. WolframAlpha ignores the word "border," and instead returns the distance from San Diego to the geographic center of Mexico. If you slightly rephrase the question about icosahedron volume by replacing the words "with edge length 2.3 meters" with "whose edges are 2.3 meters long," WolframAlpha no longer recognizes that the question is about volume; all you get back is generic information that icosahedrons have 30 edges, 20 vertices, and 12 faces, without any mention of volume. WolframAlpha

*Misunderstanding "How far is the border of Mexico from San Diego?"*

can tell you when Ella Fitzgerald was born and when she died; but if you ask it "Was Ella Fitzgerald alive in 1960?," it wrongly interprets the question as "Is Ella Fitzgerald alive?" and answers "No."

But wait, you say, what about Watson, which was so good at answering questions that it beat two human champions at *Jeopardy!* True, but unfortunately Watson is not nearly as generally powerful as it might seem. Almost 95 percent of *Jeopardy!* answers, as it turns out, are titles of Wikipedia pages. Winning at *Jeopardy!* is often just a matter of finding the right article. It's a long way from that sort of intelligent information retrieval to a system that can genuinely think and reason. Thus far, IBM hasn't even turned Watson into a robust virtual assistant. When we looked recently on IBM's web page for such a thing, all we could find was a dated demo of Watson Assistant that was focused narrowly on simulated cars, in no way on a par with the more versatile offerings from Apple, Google, Microsoft, or Amazon.

Virtual assistants like Siri and Alexa are, to be sure, starting to become useful, but they have a long way to go. And, critically, just as with Google Search, there is precious little synthesis going on. As far as we can tell, very few of them ever attempt to put together information in flexible ways from multiple sources, or even from a single source with multiple sentences, the way you did earlier when you read about Almanzo and about Ella Fitzgerald.

The truth is that *no* current AI system can duplicate what you did in those instances, integrating a series of sentences and reconstructing both what is said and what is not. If you are following what we are saying, you are human, not a machine. Someday you might be able to ask Alexa to compare *The Wall Street Journal*'s coverage of the president with *The Washington Post*'s coverage, or ask if your family doctor might have missed anything in your latest charts, but for now that's just fantasy. Better stick to asking Alexa for the weather.

What are we left with? A hodgepodge of virtual assistants, often useful, never fully reliable—not one of which can do what we humans do every time we read a book. Six decades into the history of AI, computers are still functionally illiterate.

o———o

Deep learning is not going to solve this problem, nor is the closely associated trend of "end-to-end" learning, in which an AI is trained to convert inputs directly into outputs, without any intermediate subsystems. For instance, whereas a traditional approach to driving would break things into subsystems like perception, prediction, and decision-making (perhaps using deep learning as an element in some of those subsystems), an end-to-end system would dispense with the subsystems and instead build a car-driving system that takes in camera images as input and returns, as its outputs, adjustments for acceleration and steering—without any intermediate subsystems for determining where different objects are and how they are moving, what the other drivers can be expected to do and not do, and so forth.

When it works, it can be very effective, and more straightforward to implement than more structured alternatives; end-to-end systems often require comparatively little human labor. And sometimes they are the best available solution. As the *New York Times Magazine* article on Google Translate made clear, end-to-end deep learning systems have greatly improved the state of the art in machine translation, superseding earlier approaches. Nowadays, if you want to build a program to, say, translate between French and English, you would begin by collecting an enormous corpus of documents that exist in both French and English versions, called *bitexts* (pronounced "bye-texts"), like the proceedings of the Canadian parliament, which by law must be published in both languages. From data like that, Google Translate can automatically learn the correspondences between the English words and phrases and their French counterparts, without any prior knowledge about French or English, or any prior knowledge about the intricacies of French grammar. Even skeptics like us are amazed.

The trouble is, one size doesn't fit all. Machine translation turns out to be an unusually good fit for end-to-end methods, partly because of the ready availability of large amounts of relevant data, and partly because there is generally a more-or-less clear correspondence between the English words and the French words. (Most of the time, the right French word is one of the options that you would find in a French-English dictionary, and most of the time the relation between the order of words in the two languages follows fairly standard patterns.) But many other aspects of language understanding are a much poorer fit.

Answering questions is much more open-ended, in part because the words in the correct answer to a question may bear no obvious relation to the words of the text. Meanwhile there is no database of questions and answers of a size comparable to the French-English parliamentary proceedings. Even if there were, the universe of questions and answers is so vast that any database would be but a tiny sample of all the possibilities. As explained earlier, this poses serious problems for deep learning: the further a deep learning system has to veer from its training set, the more trouble it gets into.

And, truth be told, even in machine translation, end-to-end

approaches still have their limits. They are often (but not always) fine for getting the gist, but matching words and phrases and so forth is not always enough. When getting the right translation hinges on a deeper understanding, the systems break down. If you give Google Translate the French sentence "*Je mange un avocat pour le déjeuner,*" which actually means "I eat an avocado for lunch," the translation you get is "I eat a lawyer for lunch." The French word *avocat* means both "avocado" and "lawyer," and, since people write much more often about lawyers than avocados (particularly in the proceedings of the Canadian parliament), Google Translate goes with the more frequent meaning, sacrificing sense for statistics.

In a wonderful article in *The Atlantic,* Douglas Hofstadter described the limitations of Google Translate:

> We humans know all sorts of things about couples, houses, personal possessions, pride, rivalry, jealousy, privacy, and many other intangibles that lead to such quirks as a married couple having towels embroidered "his" and "hers." Google Translate isn't familiar with such situations. Google Translate isn't familiar with situations, period. It's familiar solely with strings composed of words composed of letters. It's all about ultra-rapid processing of pieces of text, not about thinking or imagining or remembering or understanding. It doesn't even know that words stand for things.

---

For all the progress that's been made, most of the world's written knowledge remains fundamentally inaccessible, even if it is digital and online, because it is in a form that machines don't understand. Electronic health records, for example, are filled with what is often called *unstructured text,* things like doctors' notes, emails, news articles, and word-processing documents that don't fit neatly into a table. A true machine-reading system would be able to dive in, scouring doctor's notes for important information that is captured in blood tests and admission records. But the problem is so far beyond what current AI can do that many doctors' notes are never read in detail. AI tools for radiology are starting to be explored; they are able to look at images and to distinguish tumors from healthy tissue, but we have no way yet to automate another part of what a real radiologist does, which is to connect images with patient histories.

The ability to understand unstructured text is for now a significant bottleneck in a huge range of potential commercial applications of AI. We can't yet automate the process of reading legal contracts, scientific articles, or financial reports, because each consists in part of the kind of text that AI still can't grasp. Although current tools automatically pull some basic information out of even the most difficult text, a large part of the content is typically left behind. Fancier and fancier versions of text matching and link counting help—a little—but they simply don't get us to programs that can genuinely read and understand.

Of course, the situation is no better for spoken language understanding (sometimes called dialogue understanding). Even greater challenges would arise for a computerized doctor's assistant that tried to translate speech into medical notes (so that doctors could spend more time with patients and less on their laptops). Consider this simple bit of dialogue, sent to us by Dr. Vik Moharir:

> DOCTOR: Do you get chest pain with any sort of exertion?
> PATIENT: Well I was cutting the yard last week and I felt like an elephant was sitting on me. [Pointing to chest]

To a person, it's obvious that the answer to the doctor's question is "yes"; cutting the yard is in the category of exertions, and we infer that the patient experienced pain from our knowledge that elephants are heavy and being crushed by heavy things is painful. We also automatically infer that the word "felt" is being used figuratively rather than literally, given the amount of damage an actual elephant would inflict. To a machine, unless there's been a lot of specific talk of elephants before, it's probably just some rambling about large mammals and yard work.

How did we get into this mess?

Deep learning is very effective at learning correlations, such as correlations between images or sounds and labels. But deep learning struggles when it comes to understanding how objects like sentences relate to their parts (like words and phrases). Why? It's missing what linguists call compositionality: a way of constructing the meaning of a complex sentence from the meaning of its parts. For example, in the sentence *The moon is 240,000 miles from the earth*, the word *moon* means one specific astronomical object, *earth* means another, *mile* means a unit of distance, *240,000* means a number, and then, by virtue of the way that phrases and sentences work compositionally in English, *240,000 miles* means a particular length, and the sentence *The moon is 240,000 miles from the earth* asserts that the distance between the two heavenly bodies is that particular length.

Surprisingly, deep learning doesn't really have any direct way of handling compositionality; it just has lots and lots of isolated bits of information known as features, without any structure. It can learn that dogs have tails and legs, but it doesn't know how they relate to the life cycle of a dog. Deep learning doesn't recognize a dog as an animal composed of parts like a head, a tail, and four legs, or even what an animal is, let alone what a head is, and how the concept of head varies across frogs, dogs, and people, different in details yet bearing a common relation to bodies. Nor does deep learning recognize that a sentence like *The moon is 240,000 miles from the earth* contains phrases that refer to two heavenly bodies and a length.

To take another example, when we asked Google Translate to translate "The electrician whom we called to fix the telephone works on Sundays" into French, the answer we got was *L'électricien que nous avons appelé pour réparer le téléphone fonctionne le dimanche*. If you know French, you know that's not quite right. In particular, the word *works* has two translations in French: *travaille,* which means *labors,* and *fonctionne,* which means *functions properly*. Google has used the word *fonctionne,* rather than *travaille,* not grasping, as a human would, that "works on Sundays" is something that in context refers to the electrician, and that if you are talking about a person working you should be using the verb *travaille*. In grammatical terms, the subject of the verb *works* here is *electrician*, not *telephone*. The meaning of the sentence as a whole is a function of how the parts are put together, and Google doesn't really get that. Its success in many cases fools us into thinking that the system understands more than it really does, but the truth (once again illustrating the illusory progress gap) is that there is very little depth to its translations.*

A related and no less critical issue is that deep learning has no good way to incorporate background knowledge, which is something that we saw earlier, in chapter 3. If you are learning to relate an image to a label, it doesn't matter how you do it. As long as it works, nobody cares about the internal details of the system because all that matters is that you get the right label for a given image. The whole task is often relatively isolated from most of the rest of what you know.

Language is almost never like that. Virtually every sentence that we encounter requires that we make inferences about how a broad range of background knowledge interrelates with what we read. Deep learning lacks a direct way of representing that knowledge, let

* When we first wrote this sentence, in August 2018, Google Translate made the mistake that we describe. By the time we edited the draft, in March 2019, Google Translate managed to get this particular example correct. However, the fix was fragile: if you left off the period at the end or put the sentence in parentheses, or changed the sentence to "The engineer whom we called to fix the telephone works on Sundays," Google Translate reverted to its old mistake of using *fonctionne* rather than *travaille*. Because the system's behavior frequently varies, possibly from day to day, perhaps as a function of changes in the exact composition of the training data set, it is hard to guarantee that any particular sentence will or won't work from one day to the next. So long as the basic nature of the algorithm remains the same, though, the general issues we describe are likely to continue to arise.

alone performing inferences over it in the context of understanding a sentence.

And, finally, deep learning is about static translations, from an input to a label (a picture of a cat to the label *cat*), but reading is a *dynamic* process. When you use statistics to translate a story that begins *Je mange une pomme* to *I eat an apple,* you don't need to know what either sentence means, if you can recognize that in previous bitexts *je* has been matched with *I, mange* with *eat, une* with *an,* and *pomme* with *apple.*

Most of the time a machine-translation program can come up with something useful, just churning through one sentence at a time, without understanding the meaning of the passage as a whole.

When you read a story or an essay, you're doing something completely different. Your goal isn't to construct a collection of statistically plausible matches; it's to reconstruct a world that an author has tried to share with you. When you read the Almanzo story, you might first of all decide that the story contains three main characters (Almanzo, his father, and Mr. Thompson), and then you start filling in some of the details about those characters (Almanzo is a boy, his father is an adult, etc.), and you also start to try to determine some of the events that took place (Almanzo found a wallet, Almanzo asked Mr. Thompson if the wallet belonged to him, and so forth). You do something similar (largely unconsciously) every time you walk into a room, or watch a movie, or read a story. You decide what entities are there, what their relationship is to one another, and so on.

In the language of cognitive psychology, what you do when you read any text is to build up a *cognitive model* of the meaning of what the text is saying. This can be as simple as compiling what Daniel Kahneman and the late Anne Treisman called an object file—a record of an individual object and its properties—or as complex as a complete understanding of a complicated scenario.

As you read the passage from *Farmer Boy,* you gradually build up a *mental representation*—internal to your brain—of all the people and the objects and the incidents of the story and the relations among them: Almanzo, the wallet, and Mr. Thompson and also the events of Almanzo speaking to Mr. Thompson, and Mr. Thompson shouting and slapping his pocket, and Mr. Thompson snatching the wallet from Almanzo and so on. It's only after you've read the text and constructed the cognitive model that you do whatever you do with the narrative—answer questions about it, translate it into Russian, summarize it, parody it, illustrate it, or just remember it for later.

Google Translate, poster child of narrow AI that it is, sidesteps the whole process of building and using a cognitive model; it never has to reason or keep track of anything; it does what it does reasonably well, but it covers only the tiniest slice of what reading is really about. It never builds a cognitive model of the story, because it can't. You can't ask a deep learning system "what would have happened if Mr. Thompson had felt for his wallet and found a bulge where he expected to find his wallet," because it's not even part of the paradigm.

Statistics are no substitute for real-world understanding. The problem is not just that there is a random error here or there, it is that there is a fundamental mismatch between the kind of statistical analysis that suffices for translation and the cognitive model construction that would be required if systems were to actually comprehend what they are trying to read.

---

One surprisingly hard challenge for deep learning (though not for classical AI approaches) is just understanding the word *not*. Remember Siri's fail with "Find a fast food restaurant that is not McDonald's"? The person posing the query presumably wanted to get to an answer like "The Burger King at 321 Elm Street, the Wendy's at 57 Main Street, and the IHOP at 523 Spring Street." But there is nothing about Wendy's, Burger King, or IHOP that is particularly associated with the word *not*, and it doesn't happen all that frequently that someone refers to any of them as *not McDonald's,* so brute statistics don't help the way they would with relating *king* and *queen*. One can imagine some statistical tricks to solve this particular issue (identifying restaurants), but a full treatment of all the ways in which *not* can be used is way outside the scope of current approaches.

What the field really needs is a foundation of traditional computational operations, the kind of stuff that databases and classical AI are built out of: building a list (fast food restaurants in a certain neighborhood) and then excluding elements that belong on another list (the list of various McDonald's franchises).

But deep learning has been built around avoiding exactly those kinds of computations in the first place. Lists are basic and ubiquitous in computer programs and have been around for over five decades (the first major AI programming language, LISP, was literally built around lists) and yet they are not even part of the fabric of deep learning. Understanding a query with the word *not* in it thus becomes an exercise in driving square pegs into round holes.

---

And then there is the problem of ambiguity. Human languages are shot through and through with ambiguities. Words have multiple meanings: *work* (as a verb) can mean either *labors* or *functions correctly; bat* (as a noun) can mean either a flying mammal or a wooden club used in baseball. And those are comparatively clear-cut; listing all the different meanings of words like *in* or *take* fills many columns of a good dictionary. Indeed, most words except very technical ones have multiple meanings. And the grammatical structure of phrases is often ambiguous, too. Does the sentence *People can fish* mean that people are able to go fishing or that (as in Steinbeck's *Cannery Row*) people pack sardines and tuna fish into cans? Words like pronouns often introduce further ambiguities. If you say *Sam couldn't lift Harry because he was too heavy,* then, in principle, *he* could be either Sam or Harry.

What's amazing about us human readers is that 99 percent of the time, we don't even notice these ambiguities. Rather than getting confused, we quickly and with little conscious effort home in on the right way to interpret them, if there is one.*

Suppose you hear the sentence *Elsie tried to reach her aunt on the phone, but she didn't answer.* Although the sentence is logically ambiguous, there is no confusion about what it means. It does not ever consciously occur to you to wonder whether *tried* means *held court proceedings* (as in *The criminal court tried Abe Ginsburg for theft*), or whether *reach* means *physically arrive at a destination* (as in *The boat reached the shore*) or whether *on the phone* means the aunt was balanced precariously on top of the telephone (as in *a clump of dust on the phone*), or whether the word *she* in the phrase *she didn't answer* refers to Elsie herself (as it would if the sentence ended with *but she didn't get an answer*). Instead, you immediately zero in on the correct interpretation.

Now try getting a machine to do all that. In some cases, simple statistics can help. The word *tried* means *attempted* much more frequently than it means *held a court proceeding*. The phrase *on the phone* means *using the phone for communication* more frequently than it means *physically on top of the phone,* though there are exceptions. When the verb *reach* is followed by a person and the word *phone* is nearby in the sentence, it probably means *successfully established communication*.

But in many cases statistics won't get you to the right solution. Instead, there is often no way to resolve a given ambiguity without actually understanding what is going on. In the sentence that reads *Elsie tried to reach her aunt on the phone, but she didn't answer,*

* Not every ambiguity can be resolved without further information. If someone walks into the room and says *Guess what, I just saw a bat in the garage,* you really can't know whether they are talking about a flying animal or a piece of sports equipment. Until you get more context, there is nothing more that can be done, and it would not be fair to ask AI to read minds, either.

what matters is background knowledge* together with reasoning. Background knowledge makes it obvious to the reader that Elsie wouldn't answer her own phone call. Logic tells you that it must therefore be her aunt. Nobody has to teach us how to do this sort of inference in school, because we know how to do it instinctively; it follows naturally from how we interpret the world in the first place. Deep learning can't even begin to tackle this sort of problem.

o———o

Sadly, though, nothing else has really worked so far either. Classical AI techniques, of the sort that were common long before deep learning become popular, are much better at compositionality, and are a useful tool for building cognitive models, but thus far they haven't been nearly as good as deep learning at learning from data, and language is too complex to encode everything you would need strictly by hand. Classical AI systems often use templates. For example, the template [*PLACE1 is DISTANCE from PLACE2*] could be matched

* Putting this together actually requires two kinds of background knowledge. First, you need to know how telephone calls work: one person initiates the call, the other person may or may not answer; the communication is successful (the caller reaches the callee) only if the second person does answer. Second, you have to use a rule, often associated with Oxford philosopher H. P. Grice, that when people say or write things, they try to give you new information, not old information. In this case, since the sentence already said that Elsie made the call, there is no point in saying that she didn't answer it; the caller is never the person who answers a call. What is useful information is that the aunt didn't answer.

This example, by the way, is drawn from one of the most challenging tests for machines that is currently available, known as Winograd Schemas: pairs of sentences (like *Elsie tried to reach her aunt on the phone, but she didn't answer* vs. *Elsie tried to reach her aunt on the phone but she didn't get an answer*) that, at least for humans, can only be understood by making use of background knowledge. Ernie has played a central role in putting these together, along with Hector Levesque and Leora Morgenstern, and assembled a collection of Winograd Schemas online: http://www.cs.nyu.edu/faculty/davise/papers/WinogradSchemas/WS.html.

against the sentence *The moon is 240,000 miles from the earth,* and used to identify this as a sentence that specifies the distance between two places. However, each template must be hand-coded, and the minute you encounter a new sentence that differs from what comes before (say, *The moon lies roughly 240,000 miles away from the earth,* or *The moon orbits the earth at a distance of 240,000 miles*), the system starts to break down. And templates by themselves do almost nothing toward helping resolve the jigsaw puzzles of integrating knowledge of language with knowledge of the world in order to resolve ambiguity.

So far, the field of natural language understanding has fallen between two stools: one, deep learning, is fabulous at learning but poor at compositionality and the construction of cognitive models; the other, classical AI, incorporates compositionality and the construction of cognitive models, but is mediocre at best at learning.

And both are missing the main thing we have been building toward throughout this chapter: common sense.

You can't build reliable cognitive models of complex texts unless you know a lot about how the world works, about people, and places, and objects and how they interact. Without this, the vast majority of what you would read would make no sense at all. The real reason computers can't read is that they lack even a basic understanding of how the world works.

Unfortunately acquiring common sense is much harder than one might think. And, as we will see, the need for getting machines to acquire common sense is also far more pervasive than one might have imagined. If it's a pressing issue for language, it's arguably even more pressing for robotics.

# 5

# Where's Rosie?

In ten years' time Rossum's Universal Robots will be making so much wheat, so much material, so much of everything that nothing will cost anything.

—KAREL ČAPEK, WHO COINED THE WORD "ROBOT," IN *R.U.R.*, THE 1920 PLAY THAT INTRODUCED THEM

We are still in the infancy of having real autonomous interacting, learning, responsible, useful robots in our environment.

—MANUELA VELOSO, "THE INCREASINGLY FASCINATING OPPORTUNITY FOR HUMAN-ROBOT-AI INTERACTION: THE COBOT MOBILE SERVICE ROBOTS," APRIL 2018

Worried about superintelligent robots rising up and attacking us?

Don't be. At least for now, here are six things you can do in the event of a robot attack.

- Close your doors, and for good measure, lock them. Contemporary robots struggle greatly with doorknobs, sometimes even falling over as they try to open them. (In fairness, we've seen one demo that shows a robot opening one particular doorknob, in one particular lighting condition, but the AI probably doesn't generalize. We don't know of any demos that show that robots can open a broad range of doorknobs of different shapes and sizes, let alone in different lighting conditions, or any demo that shows a robot opening a locked door—even with a key.)

*Robot falling backward trying to open a door*
(Source: IEEE Spectrum)

- Still worried? Paint your doorknob black, against a black background, which will greatly reduce the chance that the robot will even be able to see it.
- For good measure, put a big poster of a school bus or a toaster on your front door (see chapter 3). Or wear a T-shirt with a picture of a cute baby. The robot will be totally confused, think you are a baby, and leave you alone.
- If that doesn't work, go upstairs, and leave a trail of banana peels and nails along the way; few robots will be able to handle an impromptu obstacle course.
- Even the ones that climb stairs probably can't hop up onto a table, unless they have been specifically trained for the task. You probably can, so hop up on the table, or climb a tree, and call 911.
- Relax; either 911 will arrive or the robot's battery will soon run out. Free-ranging robots currently typically last for a few hours, but not much more between charges, because the computers inside them demand such vast amounts of energy.

*Foiling a robot attack*

OK, this might be facetious, and maybe someday robots will be able to crash through doors and hop on tables, but for now the robots we know about are easily confused. At least in the near term, we needn't worry about Skynet, or even that robots will take our jobs.

To the contrary our biggest fear is that the robot revolution will be stillborn, because of an unreasonable fear of unlikely things.

In the movies, robots are often cast either as heroes or as demons; R2-D2 frequently rushes in to save the day; while the Terminator is here to slaughter us all. Robots want either to please their owners or to annihilate them. In the real world, robots generally don't have personalities or desires. They aren't here to slaughter us or to take our land, and they certainly don't have the wherewithal to save us from a dark lord. They don't even blip all that much, like R2-D2. Instead, for the most part, they hide on assembly lines, doing dull tasks humans would never want to do.

And robot companies, for the most part, aren't much more ambitious. One company we talked to recently is focused on building robots for excavating the foundations of buildings, another is focused on picking apples. Both seem like good business propositions, but they are not exactly the sort of stuff we dreamed of when we were kids. What we really want is Rosie, the all-purpose domestic robot from the 1960s television show *The Jetsons,* who could take care of everything in our house—the plants, the cats, the dishes, and the kids. Oh, to never have to clean anything again. But we can't buy Rosie, or anything like it, for love or for money. As we write this, there are rumors that Amazon might roll out a version of Alexa that roams around on wheels, but that's still a long way from Rosie.

The truth is that for now the bestselling robot of all time isn't a driverless car or some sort of primitive version of C-3PO, it's Roomba, that vacuum-cleaning hockey puck of modest ambition, with no hands, no feet, and remarkably little brain that we mentioned in the opening chapter, about as far from Rosie the Robot as we can possibly imagine.

To be sure, pet-like home robots are already available, and "driverless" suitcases that follow their owners around through airports may come soon enough. But the chance that a robot will be cooking and cleaning and changing kids' diapers before 2025 is practically nil. Outside of factories and warehouses, robots are still a curiosity.*

* Fast-food-cooking restaurant bots are another story; in high-volume chains like McDonald's, with great control of their internal environments and high labor costs, increased automation is likely to come soon.

What would it take to get from the modestly impressive but still greatly limited Roomba to a full-service humanoid companion like C-3PO or Rosie that could simplify almost every aspect of our domestic lives and transform the lives of the elderly and disabled, and literally save all of us hours of labor every week?

To begin with, it's important to realize that Roomba is a very different sort of creature. Inventor Rodney Brooks's great insight—inspired by reflecting on how insects with tiny brains could do complex things like flying—was that Roomba doesn't need to be very smart. Vacuuming is a mundane job and can be done decently well (not perfectly) with only a modest bit of intelligence. Even with tiny amounts of computer hardware you could make a robot that could do something useful, that people would want to spend real money on—so long as you kept the task narrow enough. If you want to pick up most of the dust, most of the time, in an ordinary room, you can just go back and forth, spiraling around and changing direction every now and then when you bump into something. It's often pretty inefficient, going over the same parts of the floor many times. But most of the time, if it doesn't miss some part of a room that can only be reached through a narrow passageway, it gets the job done.

The real challenge is to go beyond vacuuming and build robots that can carry out a broad range of complex physical tasks that we humans accomplish in daily life, from opening vacuum-sealed jars, twist-off bottle caps, and envelopes, to weeding, hedge-clipping, and mowing the lawn, to wrapping presents, painting walls, and setting the table.

Of course, there has been some progress. Our good friend the roboticist Manuela Veloso has built robots that can safely wander the halls of Carnegie Mellon University, and we have seen demos of robots lifting vastly more than their own body weight. Autonomous drones (which are flying robots) can already do some amazing things, like tracking runners as they jog along mountain trails, and (in the case of Skydio's self-flying camera) automatically avoiding trees along the way.

If you spend a few hours watching YouTube, you can see dozens of demos of robots that (at least in videos) seem vastly more powerful than Roomba. But the key word is "demo." None of it is ready for prime time. In 2016, Elon Musk announced plans to build a robotic butler, but so far as we can tell, there hasn't been much progress toward that goal. Nothing currently available commercially feels like a breakthrough, with the possible exception of the aforementioned recreational drones, which are thrilling (and tremendously useful for film crews), but not exactly Rosie either. Drones don't need to pick up things, manipulate them, or climb stairs; aside from flying around and taking photos, they aren't called on to do very much. SpotMini, a sort of headless robotic dog, is supposed to be released soon, but it remains to be seen what it will cost, and what it will be used for. Boston Dynamics' Atlas robot, a humanoid robot, about five feet tall, and 150 pounds, can do backflips and parkour, but that parkour video you saw on the web? It took twenty-one takes, in a carefully designed room; you shouldn't expect that it would be able to do the same at the playground with your kids.

Even so, there's lot of exciting hardware on the way. In addition to SpotMini and Atlas, both of which are amazing, Boston Dynamics' robots include WildCat, "the world's fastest quadruped robot," that can gallop at twenty miles per hour; and BigDog, "the First Advanced Rough-Terrain Robot," which stands three feet high and weighs 240 pounds, can run at seven miles an hour, climb slopes up to 35 degrees, walk across rubble, traverse muddy hiking trails, wade through snow and water, and carry a 100-pound payload. And of course every putative driverless car is just a robot in a car package. (And for that matter, submersibles, like Alvin, are robots, too, and so are the Mars Rovers.) Other researchers, like MIT's Sangbae Kim, are working on impressively agile hardware as well. All of this costs way too much now for the home, but someday prices will come down and robots might be in most homes.

Perhaps the most important use of robots to date has been in the shutdown and cleanup of the Fukushima nuclear reactor, after it was destroyed in the 2011 tsunami. Robots from the iRobot company were sent into the reactor to determine the state of things inside, and they continue to be used in cleaning and maintenance of the site. Though the robots were mostly controlled via radio communication by human operators outside, they also had important, though limited, AI capabilities: they could build maps, plan optimal paths, right themselves should they tumble down a slope, and retrace their path when they lost contact with their human operators.

The real issue is software. Driverless cars can propel themselves, but not safely. SpotMini is capable of amazing feats, but thus far it has been mostly teleoperated, which is to say that someone with a joystick is offstage telling the robot what to do. To be sure, the mechanical and electrical engineers and the materials scientists who make robots will be kept busy for years—there is still a long way to go in making better batteries, improving affordability, and building bodies that are sufficiently strong and dexterous—but the real bottleneck is getting robots to do what they do safely and autonomously.

What will it take to get there?

o———o

In *Star Trek: The Next Generation,* the answer is simple: all you need is what Lieutenant Commander Data has: a "positronic brain." Unfortunately, we are still not exactly sure what that is, how it might work, or where we might order one.

In the meantime, there are several things that one can expect of virtually any intelligent creature—robot, human, or animal—that aspires to be more sophisticated than Roomba. To begin with, any intelligent creature needs to compute five basic things: where it is, what is happening in the world around it, what it should do right now, how it should implement its plan, and what it should plan to do over the longer term in order to achieve the goals it has been given.

In a less-sophisticated robot focused on a single task, it might be possible to sidestep these computations to some degree. The original Roomba model had no idea of where it was, it didn't keep track of the map of the territory it had navigated, and it made no plans; it knew little more than whether or not it was moving and whether it had bumped into something recently. (More recent Roomba models do build maps, in part so that they can be more efficient, in part to make sure that they don't miss spots via an otherwise largely random search.) The question of what to do now never arose; its only goal was to vacuum.

But Roomba's elegant simplicity can only go so far. In the daily life of a more full-service domestic robot, many more choices would arise, and decision-making would thus become a more complex process—and one that would depend on having a vastly more sophisticated understanding of the world. Goals and plans could easily change from one moment to the next. A robot's owner might instruct it to unload the dishwasher, but a good domestic robot wouldn't forge ahead, no matter what; it would adapt when circumstances change.

If a glass plate falls and breaks on the floor next to the dishwasher, a robot might need to find another route to the dishwasher (a change to its short-term plans), or, better yet, it might realize that its priority is to clean up the broken glass and put the dishes on hold while it does that.

If the food on the stove catches fire, the robot has to postpone unloading the dishwasher until it has put the fire out. Poor Roomba would keep vacuuming in the middle of a Category 5 hurricane. We expect more of Rosie.

Precisely because the world changes constantly, fixed answers to the core questions about goals, plans, and the environment will never do. Instead, a high-quality domestic robot will constantly need to reevaluate. "Where am I?," "What is my current status?," "What risks and opportunities are there in my current situation?," "What should I be doing, in the near term and the long term?," and "How should I execute my plans?"* Each of these questions must be con-

* We are, of course, anthropomorphizing here, in suggesting that the robot has any sense of "I" or asks itself this kind of question. It would be more

tinuously addressed in a constant cycle, a robotic counterpart to the so-called OODA loop introduced by the legendary air force pilot and military strategist John Boyd: observe, orient, decide, and act.

The good news is that over the years, the field of robotics has gotten pretty good at implementing some parts of the robot's cognitive cycle. The bad news is that most others have seen almost no progress.

Let's start with the success stories: localization and motor control.

o———o

Localization is harder than you might think. The obvious way to start is with GPS. But GPS has until recently only been accurate to within about ten feet, and it doesn't work particularly well indoors. If that were all our hypothetical domestic robot had to work with, it could easily think it was in the bathroom when it was really on the staircase.

Military and specialized GPS can be much more accurate, but isn't likely to be available to consumer robots, which means consumer robots can't rely just on GPS. Luckily, robots can use many clues to figure out where they are, such as dead reckoning (which tracks a robot's wheels to estimate how far it has gone), vision (a bathroom looks very different from a staircase), and maps (which might be constructed in a variety of ways). Over the years, roboticists have developed a family of techniques called SLAM, short for Simultaneous Localization And Mapping, which allows robots to put together a map of their environment and to keep track of where they are in the map and where they are headed. At each step, the robot goes through the following steps:

- The robot uses its sensors to see the part of its environment that is visible from its current position.

---

accurate to write that the robot's algorithm calculates where it is, what is its current status, what are the risks and opportunities, what it should do next, and how it should execute its plans.

- It improves its current estimate of its position and orientation by matching what it is seeing against objects in its mental map.
- It adds to its mental map any objects, or parts of objects, that it has not seen before.
- It either moves (generally forward) or turns, and adjusts its estimate of its new position and orientation by taking into account how much it has moved or turned.

Although no technique is perfect, SLAM works well enough that you can plop a robot down at a random place in a well-mapped building and expect it to figure out where it is and, in conjunction with other software, how to get where it needs to go. It also allows robots to construct maps as they explore a space. Orientation, in Boyd's sense, is more or less a solved problem.

Another area with considerable progress is often called "motor control": the job of guiding a robot's motions, such as walking, lifting things, rotating its hands, turning its head, or climbing stairs.

For driverless cars, the motor control side of what needs to be done is comparatively simple. A car has only limited options, revolving around the gas pedal, the brakes, and the steering wheel. An autonomous vehicle can change its speed (or stop) and it can change its heading, by steering. There's not much else to be calculated, on the control side. Unless the car can fly, it doesn't even need to worry about the z-coordinate of going up or down in space. Computing the desired states for the steering wheel, brakes, and gas pedal from a desired trajectory is straightforward math.

The situation is way more complicated in a humanoid (or animal-like, or insect-like) robot, with multiple joints that can be moved in many ways. Suppose that there is a cup of tea on a table, and a humanoid robot is supposed to reach out its arm and grasp the cup's handle between two fingers. First, the robot has to figure out how to move the various parts of its arm and hand so that they end up at the right place without, in between, bumping into the table, having one part bump into another, or knocking the teacup over. Then it has to exert enough force on the teacup handle that it has a firm grasp but not so much force that it shatters the china. The robot must calculate a path between where it wants to go given where it is and the obstacles in its way, and then devise a complex plan (effectively a mini computer program, or a custom-built neural network) that specifies the angle at the joints and the force at the joints between body parts, and how they should change over time, perhaps as a function of feedback, in a way that never allows the contents to spill. Even just in reaching for the teacup, there might be five or more joints involved—the shoulder, the elbow, the wrist, and two fingers, with many complex interactions between them.

Despite the complexity of the problem, in recent years there has been considerable progress, most visibly at Boston Dynamics, the robot company that we mentioned earlier, which is run by Marc Raibert, a researcher with deep training in the problem of human and animal motor control. Drawing on this expertise, Raibert's robots, such as BigDog and SpotMini, move much like animals. Their software rapidly and continuously updates the forces in the actuators (the robot's "muscles") and integrates that with feedback from the robot's sensors, so that they can dynamically replan what they are supposed to do, as they are doing it (rather than just planning everything in advance and hoping for the best). Raibert's team has been able to do a bunch of things that used to be pretty challenging for many robots, like walking on uneven surfaces, climbing stairs, and even resisting forces that might otherwise knock a less stable robot down.

Lots of labs at places like Berkeley and MIT are making good progress on motor control too. YouTube has videos of laboratory demonstrations of robots that open doors, climb stairs, toss pizzas, and fold towels, although typically in carefully controlled circumstances. Although the motor control of humans remains more versatile, particularly when it comes to manipulating small objects, robots are catching up.

Then again, most of what we can glean about the current state of motor control in robotics we get from demo videos, and videos are often misleading. Often the videos have been sped up, implicitly equating what a robot could do in a minute or an hour with what a person could do in a few seconds, and sometimes relying on human operators behind the scenes. Such demo videos are proofs of concept, frequently representing the best case of something that is not truly stable, not products that are ready for shipping. They prove that a robot can, in principle, given enough time, eventually be programmed to do the physical aspects of many tasks. But they don't always tell us whether various tasks can be performed efficiently or—most important—autonomously, which of course is the ultimate goal. Eventually you should be able to say to a robot "Clean my house," and after a little bit of training, it should not only vacuum, but also dust, clean windows, sweep the porch, straighten out the books, toss the junk mail, fold the laundry, take out the garbage, and load the dishwasher. What the demos show is that we now have the hardware to do at least some of these tasks; the physical aspects of the jobs won't be the rate-limiting steps. The real challenge will be on the mental side, of having the robot correctly interpret your possibly ambiguous and vague request—relative to human goals—and coordinate all of its plans in a dynamically changing world.

As with AI more generally, the biggest challenge will be robustness. In almost any demo you watch, you see a robot doing something in the most ideal circumstances imaginable, not in a cluttered and complex environment. If you look carefully at videos of robots folding towels, you will find that the laundry is always some bright color, and the background some dark color, in an otherwise empty room, making it easier for the computer software to separate the towels from the rest of the room. In an actual home with dim light and towels that blend in with the background, pandemonium might ensue, when the robot occasionally mistakes bits of wall for bits of towel. A robotic pancake flipper might work fine in a restaurant in which it could be housed in a room with blank walls, but have trouble in a cluttered bachelor pad, where stacks of unread mail might inadvertently wind up flipped, fried, and in flames.

Motor control in the real world is not just about controlling the actions of limbs and wheels and so forth in the abstract. It's about controlling those actions relative to what an organism perceives, and coping when the world is not exactly as anticipated.

---

Situational awareness is about knowing what could happen next. Is a storm coming? Could that pot on the stove catch fire if I forget to turn it off? Could that chair be about to fall over? (Parents of toddlers tend to have heightened awareness of the latter.) One aspect of situational awareness is about looking for risk, but it can be about looking for opportunity or reward, too. For instance, a driverless car might notice that a new shortcut has opened up, or an unexpected parking spot is vacant. A home robot that was trying to unclog a drain could discover a new use for a turkey baster. On a well-controlled factory floor, situational awareness can similarly be a relatively manageable problem, limited to questions like "Is there an obstacle here?" and "Is the conveyor belt running?"

In the home, on the other hand, situations, and the risks, rewards, and opportunities that go with them, can be vastly more challenging and more complex. Sitting in your living room, you might have literally hundreds of options, and thousands of parameters could change the nature of the situation. You can get up, walk to the dining room or the kitchen, or turn on the television, or a pick up a book, or tidy the coffee table. Any of those might seem like reasonable activities on an ordinary day, but not if the smoke detector goes off or a hurricane approaches.

In calculating what's going on, and the risks and opportunities in any given moment, you (as a person) constantly combine sight with smell and hearing (and perhaps touch and taste) and a sense of where your own body is, along with an awareness of the other beings that might be in the room, your overall goals (what are you trying to do that hour? that day? that month?), and hundreds of other variables (Is it raining? Have I left a window open? Could an insect or animal wander in uninvited?). If assembly lines are closed worlds, homes are about as open-ended as you can get, and hence a serious challenge for robotics.

Driverless cars are somewhere in between. The vast majority of the time, figuring out what is going on mainly requires computing a few things: Which way am I going and how fast? Which way is the road turning? What other objects are nearby? Where are they and how are they moving? (which you can compute by comparing data from different time steps); and Where may I drive (e.g., where are the lanes, or opportunities to make a turn)? But all bets may be off in a tornado, an earthquake, or a fire, or even if there is a fender-bender or a toddler in a Halloween costume that diverts traffic.

The part of situational awareness that is fairly well handled by current AI is the job of identifying objects in some surroundings: simple object recognition is deep learning's forte. Machine-learning algorithms can now, with some degree of accuracy, identify basic elements in many scenes, from tables and pillows in a home to cars on the road. Even in simple identification, though, there are some serious problems; few object-recognition systems are robust enough to notice changes in lighting, and the more cluttered a room is, the more likely they are to grow confused. And it's not enough to note that there is a gun somewhere in the image; it is important to know whether the gun is on the wall as part of a painting (in which case it can be safely ignored), or a real object on a table, or in someone's hands pointed at someone. Even more than that, simple object-recognition systems fall far short of understanding the *relations* between objects in a scene: a mouse *in* a trap is very different from a mouse *near* a trap; a man riding a horse is very different from a man carrying a horse.

But labeling the objects in a scene is less than half the battle. The real challenge of situational awareness is to understand what all those objects collectively mean; to our knowledge there has been little or no research on this problem, which is clearly vastly harder. We don't know of any current algorithm, for example, that could look at two different scenes in which there was a fire in a living room, and reliably realize that in one case the fire is in a fireplace giving delightful warmth on a wintry day, and in the other case you had better put the fire out as fast as you can and/or call the fire department. To even approach the problem within the dominant paradigm one would probably need a bunch of labeled data sets for different kinds of homes (wooden, concrete, etc.) and different kinds of fires (grease, electrical, etc.); nobody has a general purpose system for understanding fire.

And the changing nature of the world makes situational awareness even harder; you don't want to look at the world as a snapshot, you want to see it as an unfolding movie, to distinguish objects that are toppling over from objects that are stable, and to distinguish cars pulling into a parking spot from cars pulling out of a parking spot.

Making all this even more challenging is the fact that the robot itself is both changing (for example, as it maneuvers around) and is an agent of other changes, which means that a robot must predict not only the nature of the world around it but the consequences of its own actions. On a factory floor, where everything is tightly controlled, this may be fairly easy; either the car door becomes securely attached to the car chassis, or it doesn't. In an open-ended environment, prediction becomes a real challenge: If I'm looking for the coffee, should I open the cabinet? Should I open the fridge? Should I open the mayonnaise jar? If I can't find the cover for the blender, is it OK to run the blender without it? Or to cover the blender with a plate? Even the factory floor becomes challenging the minute there is a loose screw in an unexpected place. Elon Musk blamed initial struggles in manufacturing the Tesla Model 3 on "too much automation." We suspect that a large part of the problem was that the process and environment for building the cars was dynamically changing, and the robots couldn't keep up because their programming wasn't flexible enough.

Some of this can be discovered by world experience, but the consequences of putting the cat in the blender shouldn't be something an AI learns through trial and error. The more you can make solid inferences without trying things out, the better. In this kind of everyday reasoning, humans are miles and miles ahead of anything we have ever seen in AI.

---

Perhaps an even bigger unsolved challenge lies in figuring out what is the best thing to be doing at any given moment, which is much harder (from a programming perspective) than one might initially realize.

To better understand what challenges our hypothetical domestic robot might face, let's consider three concrete scenarios, typical of what we might ask of it.

First scenario: Elon Musk is giving an evening party, and he wants a robot butler to go around serving drinks and hors d'oeuvres. For the most part this is straightforward: the robot moves around carrying plates of drinks and snacks; it collects empty glasses and plates from the guests; if a guest asks for a drink, the robot can bring it. At first blush, this might not seem far away. After all, years ago the now-defunct robot company Willow Garage had a demo of their humanoid prototype robot PR2 fetching a beer from a refrigerator.

But just as with driverless cars, true success lies in getting the details right. Real homes and real guests are complicated and unpredictable. The PR2 beer run was carefully constructed. There were no dogs, no cats, no broken bottles, and no children's toys left on the floor. Even the fridge was specially arranged, according to a colleague of ours, in ways that made the beer itself particularly accessible. But in the real world, any number of unexpected things, big and small, can go wrong. If the robot goes into a kitchen to get a wine glass and finds a cockroach in the glass, it has to form a plan that it may never have executed before, perhaps dumping the roach out of the glass, rinsing and refilling it. Or maybe the butler robot may discover a crack in the glass, in which case the glass should be safely disposed of. But what is the likelihood that some programmer is going to anticipate that exact contingency, when all of Apple's best iPhone programmers still can't even reliably automate the process of creating a calendar entry based on text in your email?

The list of contingencies is essentially endless—the Achilles' heel of narrow AI. If the robotic butler sees that a cracker has fallen on the floor, it needs to figure out how to pick up the cracker and throw it out without disturbing the guests, or it needs to be able to predict that picking up the cracker in a crowded room isn't worth the trouble, because it would cause too much commotion. No simple policy will do here, though. If the robot sees an expensive earring on the floor, rather than a cracker, the balance of the equation changes; it may be worth rescuing the earring regardless of the commotion.

Most of the time, the robot should do humans no harm. But what if a drunk guy is walking backward, oblivious to an infant crawling behind him? The robot butler ought to interfere at this point, perhaps even grabbing the drunk adult, to protect the child.

So many things might happen that they cannot possibly all be enumerated in advance, and they cannot all be found in any data set used for training. A robot butler is going to have to reason, predict, and anticipate, on its own, and it can hardly go crying to human "crowd workers" all night long every time a small decision is to be made. Surviving a night at Elon's mansion would be, from a cognitive perspective, a pretty major task.

Of course we can't all afford robotic butlers, at least not until the price comes down by a factor of a million. But now let's consider a second scenario, far less frivolous: robotic companions for the elderly and the disabled. Suppose Blake is recently blind and he would like his companion robot to help him go grocery shopping. Again, so much easier said than done, because so many kinds of things can happen. To begin with, there is basic navigation. On the way to the grocery store, Blake's companion robot will need to navigate unexpected obstacles of all kinds.

Along the way, they might encounter curbs, puddles, potholes, police, pedestrians lost in their phones, and children weaving about on scooters and skateboards. Once in the store they may need to navigate narrow aisles, or temporary taster stands that subtly alter the functional layout of the grocery store, to say nothing of the inventory people or the people cleaning the floor after somebody accidentally dropped a jar of jam. The companion robot will have to guide Blake around or through these, in addition to finding its own way. Meanwhile Blake may be accosted by an old friend, a helpful stranger, a panhandler, a policeman, a friendly dog, an unfriendly dog, or a mugger; each has to be recognized and dealt with in a different way. At the store, things have to be reached and grasped (in different ways for different items, red peppers differently from cereal boxes differently from pints of ice cream), and put into the shopping basket without cracking the eggs or piling the soup cans on top of the bananas. The shopping basket itself needs to be recognized, even though these vary in shape and size from store to store; likewise the means of payment and details of how groceries are bagged vary from one store to the next. Thousands of contingencies, varying from one shopping experience to the next, impossible to fully anticipate and program in advance.

As a third scenario, consider something like the Fukushima nuclear disaster. Imagine a building that has partially collapsed in an earthquake, and a nuclear reactor is about to melt down. A rescue robot sent inside a crisis zone has to judge what it can and cannot safely do: Can it break through a door, or cut through a wall, or does that risk further collapse? Can it safely climb a ladder that was designed for humans? If the rescue robot finds someone, what should be done? The person may be able to walk out under their own power once a path is cleared, or they may be pinned and need to be freed; or may be injured and have to be carried out carefully. If there are multiple people, the robot may have to triage, deciding which injuries should be treated first, and which not at all, given limited medical resources. If there is valuable property, the rescue robot should consider the value of the item (is it irreplaceable art?) and the urgency of removing it. All this may require deep understanding of a situation that is incompletely known, unforeseen, with features that may well be unusual or unique.

What's more, a robot needs to consider the dangers of inaction as well as action. A high-quality robot butler ought to be able to spot a Christmas tree that is listing at a dangerous angle, and readjust it, in order to keep the tree from falling over and potentially sparking and then fueling an electrical fire.

None of this is a strong point of current robots, or the AI that drives them.

So here's where things stand today, as we approach the sixty-fifth anniversary of AI: roboticists have done an excellent job of getting robots to figure out where they are, and a fairly good job of figuring how to get robots to perform individual behaviors.

But the field has made much less progress in three other areas that are essential to coping in the open-ended world: assessing situations, predicting the probable future, and deciding, dynamically, as situations change, which of the many possible actions makes the most sense in a given environment.

There is no general purpose solution either to determining what is possible and important in any given scenario, or to figuring out what a robot should do in complex and unpredictable environments. At the present time it is challenging but (with hard work) feasible to get a robot to climb a staircase or walk on uneven ground, as the Boston Dynamics prototypes have shown; it is far harder to leave a robot to clean a kitchen entirely by itself.

In a limited world, one can memorize a large number of contingencies, and interpolate between them to make guesses for unfamiliar scenarios. In a truly open-ended world, there will never be enough data. If the applesauce is growing mold, the robot needs to figure out how to respond, even if the robot has never seen anything like this before. There are just too many possibilities for it to memorize a simple table listing what to do in every circumstance that could arise.*

The real reason we don't have general-purpose domestic robots

* Programs that play games like chess and Go must, of course, deal with situations that they have not seen before, but there the kinds of situations that can arise and the choice of actions can be systematically characterized,

yet is that we don't know how to build them to be flexible enough to cope with the real world. Because the space of possibilities is both vast and open-ended, solutions that are driven purely by big data and deep learning aren't likely to suffice. Classical AI approaches, too, have been brittle, in their own ways.

All of this points, once again, to the importance of rich cognitive models and deep understanding. Even in the situation of a driverless car, a machine's internal models will need to be richer than what AI typically incorporates. Current systems are mostly limited to identifying common objects such as bicycles, pedestrians, and other moving vehicles. When other kinds of entities enter in, such limited systems can't really cope. For example, as of 2019, Tesla's Autopilot seems to have limited representations of stationary objects, such as stopped fire trucks or billboards (its first fatal accident may have been partly due to its misinterpreting a left-turning truck, with much of its mass higher than a car, as a billboard).

In the case of our household robot, the richness of the underlying cognitive model has to be considerably greater. While there are only a handful of common elements on a highway, in an average living room one might encounter chairs, a sofa or two, a coffee table, a carpet, a TV, lamps, bookcases with books, the fish tank, and the cat, plus a random assortment of children's toys. In a kitchen one might encounter utensils, appliances, cabinets, food, a faucet, a sink, more chairs and tables, the cat bowl, and again the cat. And, of course, even though kitchen utensils are usually found in the kitchen, a knife that finds its way into the living room can still hurt someone.

In many ways, what we see here echoes what we saw in the last chapter, on learning to read. Building a robot is a very different challenge from building a machine that can read, much more physical and much less about narrative and interpretation (and also much

---

and the effects of actions can be reliably predicted, in ways that do not apply in an open-ended world.

more potentially dangerous—spilling boiling tea on someone is much worse than mucking up a translation of a new story), yet we have converged on the same place.

Just as there can be no reading without rich cognitive models, there can be no safe, reliable domestic robots without rich cognitive models. Along with them, a robot will need a healthy dose of what is colloquially known as common sense: a rich understanding of the world, and how it works, and what can and cannot plausibly happen in various circumstances.

No existing AI system has all that. What sort of intelligent system does have rich cognitive models, and common sense? The human mind.

# 6

# Insights from the Human Mind

> What magical trick makes us intelligent? The trick is that there is no trick. The power of intelligence stems from our vast diversity, not from any single, perfect principle.
>
> —MARVIN MINSKY, *THE SOCIETY OF MIND*

In 2013, not long after the two of us started collaborating, we encountered a media frenzy that made our blood boil. Two researchers, Alexander Wissner-Gross and Cameron Freer, had written a paper proposing that intelligence of every kind was a manifestation of a very general physical process called "causal entropic forces." In a video, Wissner-Gross claimed to show that a system built on this idea could "walk upright, use tools, cooperate, play games, make useful social introductions, globally deploy a fleet, and even earn money trading stocks, all without being told to do so." Along with the paper, Wissner-Gross had launched an absurdly ambitious startup company called Entropica that promised "broad applications" in health care, energy, intelligence, autonomous defense, logistics, transportation, insurance, and finance.

And the media was taken in. According to the usually thoughtful science writer Philip Ball, Wissner-Gross and his co-author had "figured out a 'law' that enables inanimate objects to behave [in a way that] in effect allow[s] them to glimpse their own future. If they follow this law, they can show behavior reminiscent of some of the things humans do: for example, cooperating or using 'tools' to conduct a task." TED gave Wissner-Gross a platform to present his "new equation for intelligence."

We didn't believe a word of it, and said so, deconstructing Wissner-Gross's physics and AI, and writing rather snidely in an online piece for *The New Yorker:* "In suggesting that causal entropy might solve such a vast array of problems, Wissner-Gross and Freer are essentially promising a television set that walks your dog." In hindsight, we probably could have said the same thing in a more gentle way. But over a half decade later, there hasn't been another paper on the topic that we could find, and we don't see any sign that Wissner-Gross's mathematics of causal entropy has made any progress whatsoever. The startup company, Entropica, no longer seems active, and Wissner-Gross appears to have gone on to other projects.

Ideas like causal entropy have long seduced laypeople and scientists alike because such ideas remind us of physics: elegant, mathematical, and predictive. The media loves them because they seem like classic Big Ideas: strong statements that could conceivably alter our world, potential solutions to really complex problems in a single convenient package. Who wouldn't want to break the story about the next theory of general relativity?

The same thing happened almost a century ago in psychology, when behaviorism became all the rage; Johns Hopkins University psychologist John Watson famously claimed he could raise any child to be anything just by carefully controlling their environment, and when and where they received rewards and punishments. The premise was that what an organism would do was a straightforward mathematical function of its history. The more you are rewarded for a behavior, the more likely you are to do it; the more you are punished for it, the less likely. By the late 1950s, the psychology departments of most American universities were filled with psychologists conducting careful, quantitative experiments examining the behaviors of rats and pigeons, trying to graph everything and induce precise, mathematical causal laws.

Two decades later, behaviorism had all but disappeared, crushed by Noam Chomsky, for reasons we will discuss in a moment. What worked for the rats (in a limited set of experiments) never helped all that much in studying humans. Reward and punishment matter, but so much else does, too.

The problem, in the words of Yale cognitive scientists Chaz Firestone and Brian Scholl, is that "there is no one way the mind works, because the mind is not one thing. Instead, the mind has parts, and the different parts of the mind operate in different ways: Seeing a color works differently than planning a vacation, which works differently than understanding a sentence, moving a limb, remembering a fact, or feeling an emotion." No one equation is ever going to capture the diversity of what human minds manage to do.

Computers don't have to work in the same ways as people. There is no need for them to make the many cognitive errors that impair human thought, such as confirmation bias (ignoring data that runs against your prior theories), or to mirror the many limitations of the human mind, such as the difficulty that human beings have in memorizing a list of more than about seven items. There is no reason for machines to do math in the error-prone ways that people do. Humans are flawed in many ways, and machines need not inherit the same limitations. All the same, there is much to be learned from how human minds—which still far outstrip machines when it comes to reading and flexible thinking—work.

Here, we offer eleven clues drawn from the cognitive sciences—psychology, linguistics, and philosophy—that we think are critical, if AI is ever to become as broad and robust as human intelligence.

## 1. THERE ARE NO SILVER BULLETS.

The instant we started reading about the paper by Wissner-Gross and Freer on causal entropy, we knew it was promising too much.

Behaviorism tried to do too much as well; it was too flexible for its own good. You could explain any behavior, real or imaginary, in terms of an animal's history of reward, and if the animal did something different, you'd just emphasize a different part of that history. There were few genuine, firm predictions, just a lot of tools for "explaining" things after they happened. In the end, behaviorism really made only one assertion—one that is true and important, but too thin to be as useful as people imagined. The assertion was that animals, people included, like to do things that get rewards. This is absolutely true; other things being equal, people will choose an option that leads to a greater reward.

But that tells us too little about how, say, a person understands a line of dialogue in a film, or figures out how to use a cam lock when assembling an Ikea bookshelf. Reward is a part of the system, but it's not the system in itself. Wissner-Gross simply recast reward; in his terms, an organism is doing a good job if it resists the chaos (entropy) of the universe. None of us wants to be turned into dust, and we do resist, but that still tells us too little about how we make individual choices.

Deep learning is largely falling into the same trap, lending fresh mathematics (couched in language like "error terms" and "cost functions") to a perspective on the world that is still largely about optimizing reward, without thinking about what else needs to go into a system to achieve what we have been calling deep understanding.

But if the study of neuroscience has taught us anything, it's that the brain is enormously complex, often described as the most complex system in the known universe, and rightfully so. The average human brain has roughly 86 billion neurons, of hundreds if not thousands of different types; trillions of synapses; and hundreds of distinct proteins within each individual synapse—vast complexity at every level. There are also more than 150 distinctly identifiable brain areas, and a vast and intricate web of connections between them. As the pioneering neuroscientist Santiago Ramón y Cajal put it in his Nobel Prize address in 1906, "Unfortunately, nature seems unaware of our intellectual need for convenience and unity, and very often takes delight in complication and diversity."

Truly intelligent and flexible systems are likely to be full of complexity, much like brains. Any theory that proposes to reduce intelligence down to a single principle—or a single "master algorithm"—is bound to be barking up the wrong tree.

## 2. COGNITION MAKES EXTENSIVE USE OF INTERNAL REPRESENTATIONS.

What really killed behaviorism, more than anything else, was a book review written in 1959 by Noam Chomsky. Chomsky's target was *Verbal Behavior,* an effort to explain human language by B. F. Skinner, then one of the world's leading psychologists.

At its core, Chomsky's critique revolved around the question of whether human language could be understood strictly in terms of a history of what happened in the *external* environment surrounding the individual (what people said, and what sort of reactions they received), or whether it was important to understand the *internal* mental structure of the individual. In his conclusion, Chomsky heavily emphasized the idea that "we recognize a new item as a sentence, not because it matches some familiar item in any simple way, but because it is generated by the grammar that each individual has somehow and in some form internalized."

Only by understanding this internal grammar, Chomsky argued, would we have any hope of grasping how a child learned language. A mere history of stimulus and response would never get us there.

In its place, a new field emerged, called cognitive psychology. Where behaviorism tried to explain behavior entirely on the basis of external reward history (stimulus and response, which should remind you of the "supervised learning" that is so popular in current applications of deep learning), cognitive psychology focused largely on *internal representations,* like beliefs, desires, and goals.

What we have seen over and over in this book is the consequence of machine learning (neural networks in particular) trying to survive with too little in the way of representations. In a strict, technical sense, neural networks have representations, such as the sets of numbers known as vectors that represent their inputs and outputs and hidden units, but they are almost entirely lacking anything richer. Absent, for example, is any direct means for representing what cognitive psychologists call *propositions,* which typically describe relations between entities. For instance, in a classical AI system to represent President John F. Kennedy's famous 1963 visit to Berlin (when he said *"Ich bin ein Berliner"*), one would add a set of facts such as PART-OF (BERLIN, GERMANY), and VISITED (KENNEDY, BERLIN, JUNE 1963). Knowledge, in classical AI, consists in part of an accumulation of precisely these kinds of representations, and inference is built on that bedrock; it is trivial on that foundation to infer that Kennedy visited Germany.

Deep learning tries to fudge this, with a bunch of vectors that capture a little bit of what's going on, in a rough sort of way, but that never directly represent propositions at all. There is no specific way to represent VISITED (KENNEDY, BERLIN, JUNE 1963) or PART-OF (BERLIN, GERMANY); everything is just rough approximation. On a good day, a typical deep learning system might correctly infer that Kennedy visited Germany. But there is no reliability to it. On a bad day, that same deep learning system might get confused, and infer instead that Kennedy visited East Germany (which, of course, would have been entirely impossible in 1963) or that his brother Robert visited Bonn—because all these possibilities are nearby to one another in so-called vector space. The reason you can't count on deep learning to do inference and abstract reasoning is that it's not geared toward representing precise factual knowledge in the first place. Once your facts are fuzzy, it's really hard to get the reasoning right.

The lack of explicit representations causes similar problems for DeepMind's Atari game system. Its failure in *Breakout*—when the paddle is repositioned by a few pixels—is tightly connected with the fact that it never really comes to represent abstractions like paddles, balls, and walls, at all.

And without such representations, it's hard to build a bona fide cognitive model. And without a rich cognitive model, there can be no robustness. About all you can have instead is a lot of data, accompanied by a hope that new things won't be too different from those that have come before. But that hope is often misplaced, and when new things *are* different enough from what happens before, the system breaks down.

When it comes to building effective systems for complex problems, rich representations often turn out to be a necessity. It's no accident that when DeepMind wanted to build a system that could actually play Go at human (or superhuman) levels, they abandoned the "learn only from pixels" approach that they used in their earlier Atari game work, and started with a detailed representation of the Go board and Go rules, along with hand-crafted machinery for representing and searching a tree of moves and countermoves. As Brown University machine-learning expert Stuart Geman put it, "The fundamental challenges in neural modeling are about representations rather than learning per se."

### 3. ABSTRACTION AND GENERALIZATION PLAY AN ESSENTIAL ROLE IN COGNITION.

Much of what we know is fairly abstract. For instance, the relation "X is a sister of Y" holds between many different pairs of people: Malia Obama is a sister of Sasha Obama, Princess Anne is a sister of Prince Charles, and so on; we don't just know that a particular pair of people are sisters, we know what sisters are in general, and can apply that knowledge to individuals. We know, for example, that *if two people have the same parents, they are siblings*. If we know that Laura Ingalls Wilder was a daughter of Charles and Caroline Ingalls and then find out that Mary Ingalls was also their daughter, then we can infer that Mary was a sister of Laura's. We can also infer that it is very likely that Mary and Laura were acquainted, since most people know their siblings; that they probably had some family resemblance and some common genetic traits; and so on.

The representations that underlie both cognitive models and common sense are all built on a foundation of a rich collection of such abstract relations, combined in complex structures. Indeed, humans can abstract just about anything: pieces of time ("10:35 p.m."), pieces of space ("the North Pole"), particular events ("the assassination of Abraham Lincoln"), sociopolitical organizations ("the U.S. State Department," "the dark web"), features ("beauty," "fatigue"), relations ("sisterhood," "checkmate"), theories ("Marxism"), and theoretical constructs ("gravity," "syntax"), and use them in a sentence, an explanation, a comparison, or a story, stripping hugely complex situations down to their essentials, and giving the mind enormous leverage in reasoning broadly about the world.

Take this bit of conversation from Gary's home, which happened as we were drafting this book, at a moment when his son Alexander was five and a half years old:

ALEXANDER: What's chest-deep water?
MAMA: Chest-deep is water that comes up to your chest.
PAPA: It's different for each person. Chest-deep for me is higher than it would be for you.
ALEXANDER: Chest-deep for you is head-deep for me.

It is precisely this fluidity with inventing and extending new concepts and generalizations, often based on a tiny amount of input, that AI should be striving for.

### 4. COGNITIVE SYSTEMS ARE HIGHLY STRUCTURED.

In the bestselling book *Thinking, Fast and Slow,* Nobel laureate Daniel Kahneman divides human cognitive process into two categories, System 1 and System 2. System 1 (fast) processes are carried out quickly, often automatically. The human mind just does them; you don't have any sense of how you are doing it. When you look out at the world, you immediately understand the scene in front of you, and when you hear speech in your native language, you immediately understand what is being said. You can't control it, and you have no idea how your mind is doing it; in fact, there is no awareness that your mind is working at all. System 2 (slow) processes require conscious, step-by-step thought. When System 2 is engaged, you have an awareness of thinking: working out a puzzle, for instance, or solving a math problem, or reading slowly in a language that you are currently learning where you have to look up every third word.*

* Though it's very clear that the brain is highly structured, it's less clear exactly *how* it is structured. Nature didn't evolve the brain in order for us to

We prefer the terms *reflexive* and *deliberative* for these two systems because they are more mnemonic, but either way, it is clear that humans use different kinds of cognition for different kinds of problems. The AI pioneer Marvin Minsky went so far as to argue that we should view human cognition as a "society of mind," with dozens or hundreds of distinct "agents" each specialized for different kinds of tasks. For instance, drinking a cup of tea requires the interaction of a GRASPING agent, a BALANCING agent, a THIRST agent, and some number of MOVING agents. Howard Gardner's ideas of multiple intelligences and Robert Sternberg's triarchic theory of intelligence point in the same broad direction, as does much work in evolutionary and developmental psychology; the mind is not one thing, but many.

Neuroscience paints an even more complex picture, in which hundreds of different areas of the brain each with its own distinct function coalesce in differing patterns to perform any one computation. While the old factoid about using only 10 percent of your brain isn't true, it is true that brain activity is metabolically costly, and we rarely if ever use the entire brain at once. Instead, everything that we do requires a different subset of our brain resources, and in any given moment, some brain areas will be idle, while others are active. The occipital cortex tends to be active for vision, the cerebellum for motor coordination, and so forth. The brain is a highly structured device, and a large part of our mental prowess comes from using the right neural tools at the right time. We can expect that true artificial intelligences will likely also be highly structured, with much of their power coming from the capacity to leverage that structure in the right ways at the right time, for a given cognitive challenge.

Ironically, that's almost the opposite of the current trend. In machine learning now, there is a bias toward end-to-end models

---

dissect it. Even some of the clearest structural divisions are to some degree controversial; a version of the system 1-2 split, for example, was recently sharply critiqued by one of the originators of the theory.

that use a single homogeneous mechanism with as little internal structure as possible. An example is Nvidia's 2016 model of driving, which forsook classical divisions of modules like perception, prediction, and decision-making. Instead, it used a single, relatively uniform neural network that eschewed the usual internal divisions of labor in favor of learning more direct correlations between inputs (pixels) and one set of outputs (instructions for steering and acceleration). Fans of this sort of thing point to the virtues of "jointly" training the entire system, rather than having to train a bunch of modules (for perception, prediction, etc.) separately.

At some level such systems are conceptually simpler; one need not devise separate algorithms for perception, prediction, and the rest. What's more, at a first glance, the model appeared to work well, as an impressive video seemed to attest. Why bother with hybrid systems that treat perception and decision-making and prediction as separate modules, when it is so much easier just to have one big network and the right training set?

The problem is that such systems rarely have the flexibility that is needed. Nvidia's system worked well for hours at a time, without requiring much intervention from human drivers, but not thousands of hours (like Waymo's more modular system). And whereas Waymo's system could navigate from point A to point B and deal with things like lane changes, all Nvidia's could do was to stick to a lane; important, but just a tiny part of what is involved in driving. (These systems are also harder to debug, as we will discuss later.)

When push comes to shove and the best AI researchers want to solve complex problems, they often use hybrid systems, and we expect this to become more and more the case. DeepMind was able to solve Atari games (to some degree) without a hybrid system, training end-to-end from pixels and game score to joystick actions, but could not get a similar approach to work for Go, which is in many ways more complex than the low-resolution Atari games from the 1970s and 1980s. There are, for instance, vastly more possible game positions, and actions can have much more intricate consequences in Go. Bye-bye pure end-to-end systems, hello hybrids.

Achieving victory in Go required putting together two different approaches: deep learning and a second technique, known as Monte Carlo Tree Search, for sampling possibilities among a branching tree of possible ways of continuing in a game. Monte Carlo Tree Search is itself a hybrid of two other ideas, both dating from the 1950s: game tree search, a textbook AI technique for looking forward through the players' possible future moves, and Monte Carlo search, a common method for running multiple random simulations and doing statistics on the results. Neither system on its own—deep learning or Monte Carlo Tree Search—would have produced a world champion. The lesson here is that AI, like the mind, must be structured, with different kinds of tools for different aspects of complex problems.*

## 5. EVEN APPARENTLY SIMPLE ASPECTS OF COGNITION SOMETIMES REQUIRE MULTIPLE TOOLS.

Even at a fine-grained scale, cognitive machinery often turns out to be composed not of a single mechanism, but many.

Take verbs and their past tense forms, a mundane-seeming system, which Steven Pinker once called the fruit flies of linguistics: simple "model organisms" from which much can be learned. In English and

* In a sort of terminological imperialism, advocates of deep learning often refer to a system, no matter how complex, that contains *any* deep learning within, as a deep learning system, no matter what role deep learning might play in the larger system, even if other, more traditional elements play a critical role. To us, this seems like calling a car a transmission, just because a transmission plays an important role in the car, or a person a kidney, just because they couldn't live without at least one. Kidneys are obviously critical for human biology, but it doesn't mean that the study of medicine should be reconstrued as nephrology writ large. We anticipate that deep learning will play an important role in hybrid AI systems, but that doesn't mean that they will rely exclusively or even largely on deep learning. Deep learning is much more likely to be a necessary component of intelligence than to be sufficient for intelligence.

many other languages, some verbs form their past tense regularly, by means of a simple rule (*walk-walked, talk-talked, perambulate-perambulated*), while others form their past tense irregularly (*sing-sang, ring-rang, bring-brought, go-went*). Part of Gary's PhD work with Pinker focused on children's overregularization errors (in which an irregular verb is inflected as if it were a regular verb, such as *breaked* and *goed*). Based on the data they analyzed, they argued for a hybrid model, a tiny bit of structure at the micro level, in which regular verbs were generalized by rules (much as one might find in computer programs and classical AI), whereas irregular verbs were produced through an associative network (which were basically predecessors of deep learning). These two different systems co-exist and complement each other; irregulars leverage memory, regulars generalize even when few directly relevant pieces of data are available.

Likewise, the mind deals with concepts in a number of different modes; partly by definitions, partly by typical features, partly by key examples. We often simultaneously track what is typical of a category and what must be true of it in order for it to meet some formal criteria. Grandmother Tina Turner danced about in miniskirts. She may not have looked like a typical grandmother, but she met the relational criteria just fine: she had children, and those children in turn had children.

A key challenge for AI is to find a comparable balance, between mechanisms that capture abstract truths (most mammals bear live young) and mechanisms that deal with the gritty world of exceptions (the platypus lays eggs). General intelligence will require both mechanisms like deep learning for recognizing images and machinery for handling reasoning and generalization, closer to the mechanisms of classical AI and the world of rules and abstraction.

As Demis Hassabis recently put it, "true intelligence is a lot more than just [the kind of perceptual classification deep learning has excelled at], you have to recombine it into higher-level thinking and symbolic reasoning, a lot of the things classical AI tried to deal with in the 80s." Getting to broad intelligence will require us to bring together many different tools, some old, some new, in ways we have yet to discover.

## 6. HUMAN THOUGHT AND LANGUAGE ARE COMPOSITIONAL.

The essence of language, for Chomsky, is, in a phrase from an earlier linguist, Wilhelm von Humboldt (1767–1835), "infinite use of finite means." With a finite brain and finite amount of linguistic data, we manage to create a grammar that allows us to say and understand an infinite range of sentences, in many cases by constructing larger sentences (like this one) out of smaller components, such as individual words and phrases. If we can say *the sailor loved the girl,* we can use that sentence as a constituent in a larger sentence (*Maria imagined that the sailor loved the girl*), which can serve as a constituent in a still larger sentence (*Chris wrote an essay about how Maria imagined that the sailor loved the girl*), and so on, each of which we can readily interpret.

At the opposite pole is the pioneering neural network researcher Geoff Hinton, every bit as much a leader in his world as Chomsky has been in linguistics. Of late, Hinton has been arguing for what he calls "thought vectors." A vector is just a string of numbers like [40.7128° N, 74.0060° W], which is the longitude and latitude of New York City, or [52,419, 663,268, . . . 24,230, 97,914] which are the areas in square miles of the U.S. states in alphabetical order. In deep learning systems, every input and every output can be described as a vector, with each "neuron" in the network contributing one number to the relevant vector. As a result, people in the machine-learning world have tried to encode words as vectors for a number of years, with the notion that any two words that are similar in meaning ought to be encoded with similar vectors. If *cat* is encoded as [0, 1, -0.3, 0.3], perhaps dog would be encoded as [0, 1, -0.35, 0.25]. A technique called Word2Vec, devised by Ilya Sutskever and Tomas Mikolov when they were at Google, allowed computers to efficiently and quickly come up with word vectors of this sort, each one made up of a couple hundred real numbers, based on the other words that tend to appear nearby it in texts.*

In certain contexts the technique works well. Take the word *saxophone*. Across a large collection of written English, *saxophone* occurs near words like *play* and *music* and names like *John Coltrane* and *Kenny G*. Across a large database, the statistics for *saxophone* are close to the statistics for *trumpet* and *clarinet* and far from the statistics for *elevator* and *insurance*. Search engines can use this technique (or minor variations on it) to identify synonyms; product search on Amazon has also become much better thanks to techniques like these.

What really made Word2Vec famous, though, was the discovery that it seemed to work for verbal analogies, like *man is to woman as king is to* __. If you add together the numbers representing *king* and *woman* and subtract the numbers for the word *man*, and then you look for the nearest vector, presto, you get the answer *queen*, without any explicit representation anywhere of what a king is, or what a woman is. Where traditional AI researchers spent years trying to define those notions, Word2Vec appeared to have cut the Gordian knot.

Based in part on results like these, Hinton sought to generalize the idea. Instead of representing sentences and thoughts in terms of complex trees, which interact poorly with neural networks, why not represent thoughts as vectors? "If you take the vector for *Paris* and subtract the vector for *France* and add *Italy*, you get *Rome*," Hinton told *The Guardian*. "It's quite remarkable." Similar techniques, as Hinton has pointed out, underlie Google's recent advances in machine translation; why not represent all thoughts this way?

Because sentences are different from words. You can approximate

* A variety of other techniques have been used for encoding words as vectors, an approach that is broadly called *embedding*. Some are more complex, some simpler but more efficient for computers to calculate; each yields slightly different results, but the fundamental limitations are similar.

the meaning of a word by considering how it's been used across a bunch of different circumstances; the meaning of *cat* is something at least a little like the average of all the uses of *cat* you have heard before, or (more technically) like the cloud of points in a vector space that a deep learning system uses to represent it. But every sentence is different; *John is easy to please* isn't all that similar to *John is eager to please,* even though the letters in the two sentences aren't that different. And *John is easy to please* is very different from *John is not easy to please;* adding a single word can change the meaning altogether.

The ideas and the nuanced relationships between them are just way too complex to capture by simply grouping together sentences that ostensibly seem similar. We can distinguish between the phrase *a book that is on the table* and the phrase *a table that is on a book,* and both from the phrase *the book that is not on the table,* and each of these from the sentence *Geoffrey knows that Fred doesn't give a whit about the book that is on the table, but that he does care a lot about the large and peculiar sculpture of a fish that currently has a table balanced on top of it, particularly since the table is listing to the right and might fall over at any second.* Each of these sentences can be endlessly multiplied, each with distinct meanings; in each case, the whole is quite distinct from statistical averages of its parts.*

It is precisely for this reason that linguists typically represent

* Indeed, even just going back to words for a second, there are serious problems in trying to map complex concepts onto vectors. If the arithmetic of subtracting man from the sum of king and woman happens to work well in that particular instance, the system of translating words into vectors is hardly robust overall. Given the analogy *short is to tall as beautiful is to____*, the top five answers of the Word2Vec system are *tall, gorgeous, lovely, stunningly beautiful,* and *majestic,* rather than *ugly. Lightbulb is to light as radio is to_____* yields *light, FM, Radio,* and *radio_station* rather than *sound* or *music.* And, troublingly, *moral* is judged to be closer to *immoral* than to *good.* For all the hype, the reality is that Word2Vec doesn't grasp even basic notions like opposites.

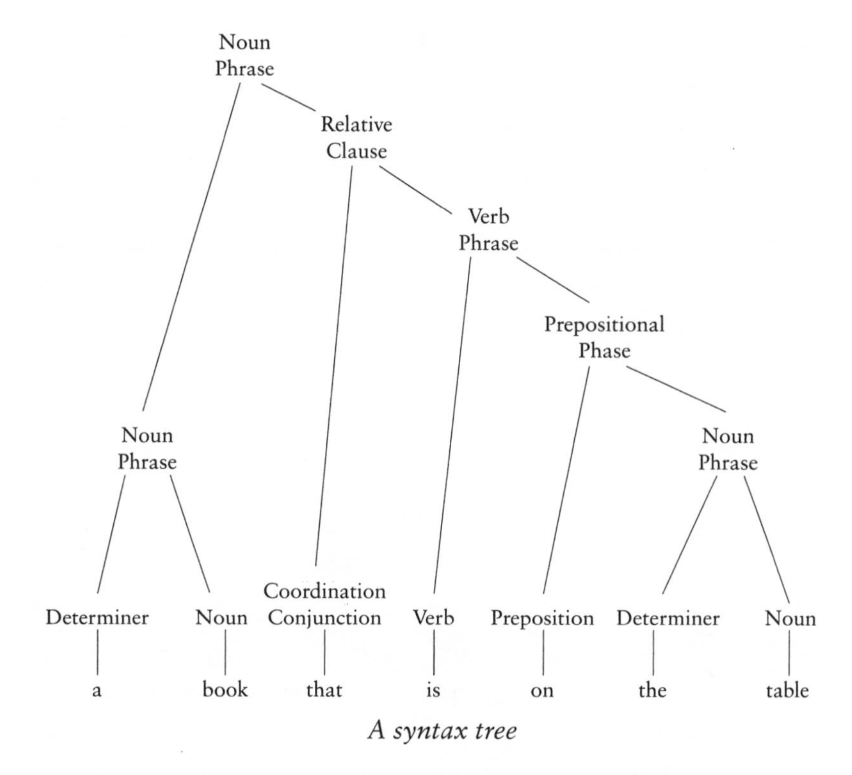


*A syntax tree*

language with branching diagrams called trees (usually drawn with the root at the top).

In this framework, each component of a sentence has its place, and it is easy to distinguish one sentence from the next, and to determine the relations between those elements, even if two sentences share most or all of their words. In working without such highly structured representations of sentences, deep learning systems tend to get themselves in trouble in dealing with subtleties.

A deep-learning-powered "sentiment analyzer," for example, is a system that tries to classify whether sentences are positive or negative. In technical terms, each sentence is transformed into a vector, and the presumption is that positive sentences ("Loved it!") will be represented with one set of vectors that are similar to one another ("cluster together"), and negative sentences ("Hated it!") will be represented by another set of vectors that group together in a separate cluster. When a new sentence comes along, the system essentially figures out whether it is closer to the set of positive vectors or the set of negative vectors.

Many input sentences are obvious, and classified correctly, but subtle distinctions are often lost. Such systems cannot distinguish between, say, "Loved it until I hated it" (a negative review about a film gone wrong) and "Hated it until I loved it" (a more positive review about a film that gets off to a slow start before redeeming itself), because they don't analyze the structure of a sentence in terms of how it relates to its component parts—and, critically, they don't understand how the meaning of a sentence derives from its parts.

The moral is this: statistics often approximate meaning, but they never capture the real thing. If they can't capture individual words with precision, they certainly aren't going to capture complex thoughts (or the sentences that describe them) with adequate precision. As Ray Mooney, computational linguist at the University of Texas at Austin, once put it, profanely, but not altogether inaccurately, "You can't cram the meaning of an entire fucking sentence into a single fucking vector!" It's just too much to ask.*

### 7. A ROBUST UNDERSTANDING OF THE WORLD REQUIRES BOTH TOP-DOWN AND BOTTOM-UP INFORMATION.

Take a look at this picture. Is it a letter or a number?

*Letter B or numeral 13?*

Quite obviously, it could be either, depending on the context.

* Strictly speaking, it could be done, using techniques such as Gödel numbering, which maps every sentence onto a number that is calculated in a highly structured way, but doing so is a kind of Pyrrhic victory that would

*Interpretation depends on context.*

Cognitive psychologists often distinguish between two kinds of knowledge: bottom-up information, which is information that comes directly from our senses, and top-down knowledge, which is our prior knowledge about the world (for example, that letters and numbers form distinct categories, that words and numbers are composed from elements drawn from those categories, and so forth). An ambiguous image looks one way in one context and another in a different context because we try to integrate the light falling on our retina with a coherent picture of the world.

Read a psychology textbook, and you will see dozens of examples. One classic experiment asked people to look at pictures like this and then draw them from memory, after first priming them with particular phrases, like *sun* or *ship's wheel* for the one at the bottom, *curtains in a window* or *diamond in a rectangle* for the one on top.

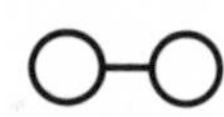

*Images with multiple interpretations*

---

require giving up the very numerical similarity between similar sentences that systems driven by backpropagation rely on.

| Some reproductions | Label list 1 | Original stimuli | Label list 2 | Some reproductions |
|---|---|---|---|---|
| | Curtains in a window | | Diamond in a rectangle | |
| | Crescent moon | | Letter "C" | |
| | Eyeglasses | | Dumbbells | |
| | Seven | | Four | |
| | Ship's wheel | | Sun | |

*How a picture is reconstructed depends on context.*

How people reconstructed the pictures depended heavily on the labels they were given.

One of our personal favorite demonstrations about the importance of context in perception comes from Antonio Torralba's lab at MIT, which showed a picture of a lake containing a ripple that was shaped vaguely like a car, enough so that a vision system was fooled. If you zoomed in on the ripple in enough detail, you indeed find smears of light that looked car-like, but no human would be fooled, because we know cars can't actually travel on lakes.

To take another example, look at the details that we have extracted from this picture of Julia Child's kitchen (opposite).

Can you identify the fragments below it? Of course you can; the picture on the left is the kitchen table with two chairs placed around it (the top of a third chair on the far side is a just barely visible crescent) and a plate on a placemat on top. The picture on the right is just the chair to the left of the table.

But the pixels of the table and the chairs alone don't tell us this.

*Julia Child's kitchen*

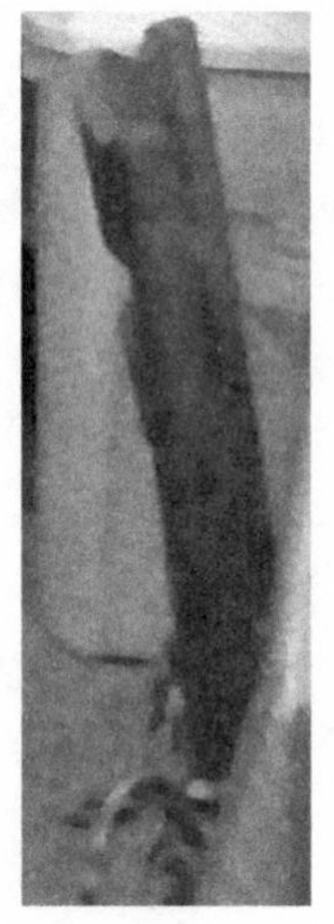

*Details from kitchen image*

If we run these by themselves through Amazon's photo-detection software (Rekognition), it labels the left-hand picture as "Plywood" with 65.5 percent confidence, and the right hand as "Dirt Road" or "Gravel" with 51.1 percent confidence. Without context, the pixels on their own make little sense.

Much the same applies in our understanding of language. One area where context helps is the resolution of ambiguity, mentioned earlier. When one of us saw a sign the other day on a country road reading "free horse manure," it could, logically speaking, have been a request (with the same grammar as Free Nelson Mandela) or a giveaway of something an owner had too much of; there was no trouble telling which, because horse manure doesn't yearn to be free.*

World knowledge is also critical for interpreting nonliteral language. When one restaurant server tells another, "The roast beef wants some coffee," nobody thinks that a roast beef sandwich has suddenly become thirsty; we infer that the *person* who ordered the roast beef wants a beverage; what we know about the world tells us that sandwiches themselves don't have beliefs or desires.

In the technical terminology of linguistics, language tends to be underspecified, which means we don't say everything we mean; we leave most of it to context, because it would take *forever* to spell everything out.

Top-down knowledge affects our moral judgments, too. Most people, for example, think that killing is wrong; yet many will make exceptions for war, self-defense, and vengeance. If I tell you in isolation that John Doe killed Tom Tenacious, you will think it's wrong. But if you see Doe kill Tenacious in the context of a Hollywood movie in which Tenacious first capriciously killed Doe's family, you will probably cheer when Doe pulls the trigger in vengeance. Stealing is wrong, but Robin Hood is cool. How we understand things is rarely a pure matter of bottom-up data (a murder or theft took place) in isolation, but always a mixture of that data and more abstract, higher-level principles. Finding a way to integrate the two—bottom-up and top-down—is an urgent but oft-neglected priority for AI.

* For that matter, it could also have been the manure of horses that were given away for free, another possibility that our brains are clever enough to automatically disregard.

## 8. CONCEPTS ARE EMBEDDED IN THEORIES.

According to Wikipedia, a quarter (as a unit of American currency) is "a United States coin worth 25 cents, about an inch in diameter." A pizza is "a savoury dish of Italian origin." Most are circles, a few are rectangles, and still others, less common, are ovals or other shapes. The circles typically range in size from six to eighteen inches inches in diameter. Yet, as Northwestern University cognitive psychologist Lance Rips once pointed out, you can easily imagine (and might be willing to eat, perhaps as a cutesy appetizer) a pizza whose diameter was precisely that of an American quarter. On the other hand, you would never accept as legitimate currency a facsimile of a quarter that was even 50 percent larger than a standard quarter; instead, you would dismiss it as a poor quality counterfeit.

*Paying for a quarter-sized pizza with a pizza-sized quarter*

This is in part because you have different intuitive theories of money and food. Your theory of money tells you that we are willing to trade physical things of value (like food) for markers (like coins and bills) that denote abstract value, but that exchange rests on the markers having legitimacy. Part of that legitimacy rests on those markers being issued by some special authority, such as a mint, and part of the way we assess that legitimacy is that we expect the markers to fulfill exact requirements. Thus a quarter can't be the size of a pizza.

At one point, psychologists and philosophers tried to define concepts strictly in terms of "necessary and sufficient" conditions; a square must have four equal sides, connected with 90-degree angles; a line is the shortest distance between points. Anything that meets the criteria qualifies, anything outside them doesn't; if two sides are unequal, you no longer have a square. But scholars struggled trying to define concepts that were less mathematical. It's hard to give exact criteria for a bird or a chair.

Another approach involved looking at particular instances, either a central case (a robin is a prototypical bird) or a set of examples (for instance, all the birds you might have seen). Since the 1980s, many have favored a view, which we share, in which concepts are embedded in theories. Our brains seem to do a decent job of tracking both individual examples and prototypes, but we can also reason about concepts relative to the theories they are embedded in, as in the example of pizzas and quarters. To take another example, we can understand a biological creature as having a "hidden essence" independent of virtually all of its perceptual properties.

In a classic experiment, the Yale psychologist Frank Keil asked children whether a raccoon that underwent cosmetic surgery to look like a skunk, complete with "super smelly" stuff embedded, could become a skunk. The children were convinced that the raccoon would remain a raccoon nonetheless, despite perceptual appearances and functional properties like scent, presumably as a consequence of their theory of biology, and the notion that it's what is inside a creature that really matters. (An important control showed that the children didn't extend the same theory to human-made artifacts, such as a coffeepot that was modified through metalworking to become a bird feeder.)

We see concepts that are embedded in theories as vital to effective learning. Suppose that a preschooler sees a photograph of an iguana for the first time. Almost immediately, the child will be able to recognize not only other photographs of iguanas, but also iguanas in videos and iguanas in real life, with reasonable accuracy, easily distinguishing them from kangaroos and maybe even from some other lizards. Likewise, the child will be able to infer from general knowledge about animals that iguanas eat and breathe; that they are born small, grow, breed, and die; and that there is likely to be a population of iguanas, all of whom look more or less similar and behave in similar ways.

No fact is an island. To succeed, a general intelligence will need to embed the facts that it acquires into richer overarching theories that help organize those facts.

## 9. CAUSAL RELATIONS ARE A FUNDAMENTAL ASPECT OF UNDERSTANDING THE WORLD.

As Turing Award winner Judea Pearl has emphasized, a rich understanding of causality is a ubiquitous and indispensable aspect of human cognition. If the world was simple, and we had full knowledge of everything, perhaps the only causality we would need would be physics. We could determine what affects what by running simulations; if I apply a force of so many micronewtons, what will happen next?

But as we will discuss in detail below, that sort of detailed simulation is often unrealistic; there are too many particles to track in the real world, and too little time.

Instead, we often use approximations; we know things are causally related, even if we don't know exactly why. We take aspirin, because we know it makes us feel better; we don't need to understand the biochemistry. Most adults know that having sex can lead to babies, even if they don't understand the exact mechanics of embryogenesis, and can act on that knowledge even if it is incomplete. You don't have to be a doctor to know that vitamin C can prevent scurvy, or a mechanical engineer to know that pressing a gas pedal makes the car go faster. Causal knowledge is everywhere, and it underlies much of what we do.

In Lawrence Kasdan's classic film *The Big Chill*, Jeff Goldblum's character jokes that rationalizations are even more important than sex. ("Have you ever gone a week without a rationalization?" he asks.) Causal inferences are even more important than rationalizations; without them, we just wouldn't understand the world, even for an hour. We are with Pearl in thinking that few topics in AI could be more important; perhaps nothing else so important has been so neglected. Pearl himself has developed a powerful mathematical theory, but there is much more to explore about how we manage to learn the many causal relationships that we know.

That's a particularly thorny problem, because the most obvious route to causal knowledge is so fraught with trouble. Almost every cause that we know leads to correlations (cars really do tend to go faster when you press the gas pedal, so long as the engine is running and the emergency brake hasn't been deployed), but a lot of correlations are not in fact causal. A rooster's crow reliably precedes dawn; but any human should be able to tell you that silencing a rooster will not stop the sun from rising. The reading on a barometer is closely correlated with the air pressure, but manually moving a barometer needle with your hands will not change the air pressure.

Given enough time, it's easy to find all kinds of purely coincidental correlations, like this one from Tyler Vigen, correlating per capita cheese consumption and death by bedsheet tangling, sampled from the years 2000–2009.

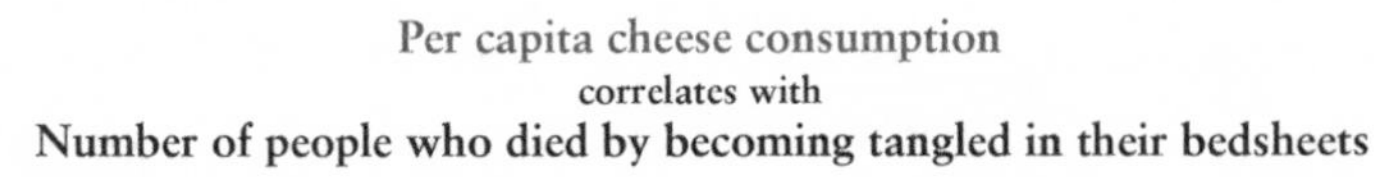


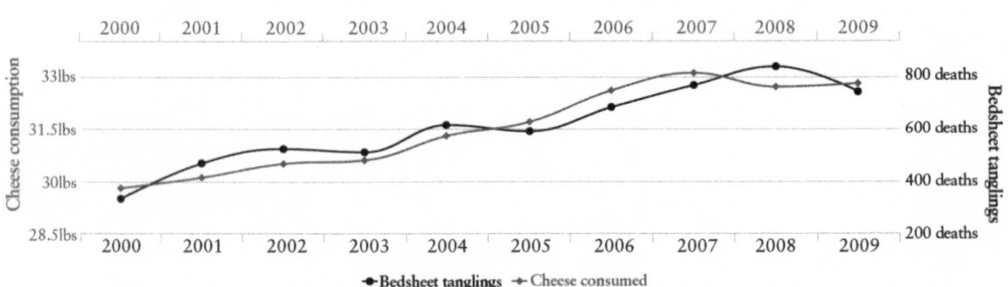


*A spurious correlation*

As a law student, Vigen compiled a whole book of these, entitled *Spurious Correlations*. In the same time period, Vigen notes, the number of people who drowned by falling into a pool was closely correlated with the number of films that Nicolas Cage appeared in. Getting a machine to realize that these sorts of tongue-in-cheek correlations are spurious, with no genuine causal connection, but that the correlation between gas pedal and acceleration is instead a genuine causal relation, will be a major accomplishment.*

## 10. WE KEEP TRACK OF INDIVIDUAL PEOPLE AND THINGS.

As you go through daily life, you keep track of all kinds of individual objects, their properties, and their histories. Your spouse used to work as a journalist and prefers brandy to whiskey. Your daughter used to be afraid of thunderstorms and she prefers ice cream to cookies. Your car has a dent on the back right door, and you got the transmission replaced a year ago. The drugstore at the corner used to sell high-quality merchandise, but it has gone downhill since it was sold to new management. Our world of experience is made up of individual things that persist and change over time, and a lot of what we know is organized around those things, not just cars and people and drugstores in general, but particular entities and their individual histories and idiosyncrasies.

Strangely, that's not a point of view that comes at all naturally to a deep learning system. Deep learning is focused around categories, not around individuals. For the most part, deep learning systems are good at learning generalizations: children mostly prefer dessert to vegetables, cars have four wheels, and so on. Those are the kinds of

* We are not suggesting this is easy for humans. It took decades before many people were willing to accept the fact that smoking increases the risk of lung cancer. For all the nineteenth century much of the medical establishment fiercely resisted the idea that puerperal fever was being spread by doctors with unsterilized hands, not merely because it hurt their professional pride, but also because they considered that it made no sense. The doctors were washing their hands with soap; how could the tiny amounts of contaminant that might remain be so destructive?

facts that deep learning systems find natural, not facts specifically about *your* daughter and *your* car.

Certainly, there are exceptions, but if you look carefully, those are exceptions that prove the rule. For example, deep learning systems are very good at learning to identify pictures of individual people; you can train a deep learning system to recognize pictures of Derek Jeter, say, with high accuracy. But that's because the system thinks of "pictures of Derek Jeter" as a category of similar pictures, not because it has any idea of Derek Jeter as an athlete or an individual human being. The deep learning mechanism for learning to recognize an individual like Derek Jeter and learning to recognize a category, like baseball player, is essentially the same: they are both categories of images. It is easier to train a deep learning system to recognize pictures of Derek Jeter than to get it to infer from a set of news stories over a series of years that he played shortstop for the Yankees from 1995 to 2014.

Likewise, it is possible to get a deep learning system to track a person in a video with some accuracy. But to the deep learning system, that's an association of one patch of pixels in one video frame to another patch of pixels in the next frame; it doesn't have any deeper sense of the individual behind it. It has no idea that when the person is not in the video frame, he or she is still somewhere. It would see no reason to be surprised if one person stepped into a phone booth and then two people came out.

## 11. COMPLEX COGNITIVE CREATURES AREN'T BLANK SLATES.

By 1865 Gregor Mendel had discovered that the core of heredity was what he called a factor, or what we now call a gene. What he didn't know was what genes were made of. It took almost eighty years until scientists found the answer. For decades, many scientists went down a blind alley, mistakenly imagining that Mendel's genes were made out of some sort of protein; few even imagined they were made out of a lowly nucleic acid. Only in 1944 did Oswald Avery use the process of elimination to finally discover the vital role of DNA. Even then, most people paid little attention because at that time, the scientific community simply wasn't "very interested in nucleic acids." Mendel himself was initially ignored, too, until his laws were rediscovered in 1900.

Contemporary AI may be similarly missing the boat when it comes to the age-old question of innateness. In the natural world, the question is often phrased as "nature versus nurture." How much of the structure of the mind is built in, and how much of it is learned? Parallel questions arise for AI: Should everything be built in? Learned?

As anyone who has seriously considered the question will realize, this is a bit of a false dichotomy. The evidence from biology—from fields like developmental psychology (which studies the development of babies) and developmental neuroscience (which nowadays studies the relation between genes and brain development)—is overwhelming: nature and nurture work together, not in opposition. Individual genes are in fact levers of this cooperation, as Gary emphasized in his book *The Birth of the Mind*. (As he noted there, each gene is something like an "IF-THEN" statement in a computer program. The THEN side specifies a particular protein to be built, but that protein is only built IF certain chemical signals are available, with each gene having its own unique IF conditions. The result is like an adaptive yet highly compressed set of computer programs, executed autonomously by individual cells, in response to their environments. Learning itself emerges from this stew.)

Strangely, the majority of researchers in machine learning don't seem to want to engage with this aspect of the biological world.*

* An unusual fact about human beings may contribute to a widespread bias against innateness. Because human infants' heads are exceptionally large relative to the size of the birth canal, we are born before our brains are fully developed (unlike many precocial animals that walk from the moment of birth). The brain likely continues to physically develop and mature endogenously, in part independently of experience after birth, in the same way that facial hair doesn't appear until after puberty. Not everything that happens in the first few months of life is learned, but people often attribute virtually every postnatal developmental change to experience, thereby overestimating

Articles on machine learning rarely make any contact with the vast literature in developmental psychology, and when they do, it is sometimes only to reference Jean Piaget, who was an acknowledged pioneer in the field but died nearly forty years ago. Piaget's questions—for example, "Does a baby know that objects continue to exist even after they have been hidden?"—still seem right on target, but the answers he proposed, like his theory of stages of cognitive development and his guesses about the ages at which children discovered things, were based on obsolete methodologies that haven't stood the test of time; they are now an outdated reference point.

It is rare to see any research in developmental psychology from the last two decades cited in the machine-learning literature and even rarer to see anything about genetics or developmental neuroscience cited. Machine-learning people, for the most part, emphasize learning, but fail to consider innate knowledge. It's as if they think that because they study learning, nothing of great value could be innate. But nature and nurture don't really compete in that way; if anything, the richer your starting point, the more you can learn. Yet for the most part, deep learning is dominated by a "blank slate" perspective that is far too dismissive of any sort of important prior knowledge.*

We expect that in hindsight this will be viewed as a colossal oversight. We don't, of course, deny the importance of learning from experience, which is obvious even to those of us who value innateness. However, learning from an absolutely blank slate, as machine-learning researchers often seem to wish to do, makes the game much harder than it should be. It's nurture without nature, when the most effective solution is obviously to combine the two.

---

the importance of learning and underestimating the importance of genetic factors.

* Strictly speaking, no system is entirely without some sort of innate structure. Every deep learning system, for example, is innately endowed by its programmers with a specific number of layers, a specific pattern of interconnectivity between nodes, specific mathematical functions for acting on inputs to those nodes, specific rules for learning, specific schemes for what input and output units stand for, and so forth.

As Harvard developmental psychologist Elizabeth Spelke has argued, humans are likely born understanding that the world consists of enduring objects that travel on connected paths in space and time, with a sense of geometry and quantity, and the underpinnings of an intuitive psychology. Or, as Kant argued two centuries earlier, in a philosophical context, an innate "spatiotemporal manifold" is indispensable if one is to properly conceive of the world.

It also seems very likely that some aspects of language are also partly prewired innately. Children may be born knowing that the sounds or gestures the people around them make are communications that carry meaning; and this knowledge connects to other innate, basic knowledge of human relations (Mommy will take care of me, and so on). Other aspects of language may be innate as well, such as the division of language into sentences and words, expectations about what language might sound like, the fact that a language has a syntactic structure, and the fact that syntactic structures relate to a semantic structure.

In contrast, a pure blank-slate learner, which confronts the world as a pure audiovisual stream, like an MPEG4 file, would have to learn everything, even the existence of distinct, persistent people. A few people have tried to do something like this, including at DeepMind, and the results haven't been nearly as impressive as the same approach applied to board games.

Within the field of machine learning, many see wiring anything innately as tantamount to cheating, and are more impressed with solutions if there is as little as possible built in. Much of DeepMind's best-known early work seemed to be guided by this notion. Their system for playing Atari games built in virtually nothing, other than a general architecture for deep reinforcement learning, and features that represented joystick options, screen pixels, and overall score; even the game rules themselves had to be induced from experience, along with every aspect of strategy.

In a later paper in *Nature*, DeepMind alleged that they had mastered Go "without human knowledge." Although DeepMind certainly used less human knowledge about Go than their predecessors, the phrase "without human knowledge" (used in their title) overstated the case: the system still relied heavily on things that human researchers had discovered over the last few decades about how to get machines to play games like Go, most notably Monte Carlo Tree Search, the way described earlier of randomly sampling from a tree of different game possibilities, which has nothing intrinsic to do with deep learning. DeepMind also (unlike in their earlier, widely discussed work on Atari games) built in the rules and some other detailed knowledge about the game. The claim that human knowledge wasn't involved simply wasn't factually accurate.

More than that, and just as important, the claim itself was revealing about what the field values: an effort to eliminate prior knowledge, as opposed to an effort to leverage that knowledge. It would be as if car manufacturers thought it was cool to rediscover round wheels, rather than to just use wheels in the first place, given the vast experience of two millennia of previous vehicle building.

The real advance in AI, we believe, will start with an understanding of what kinds of knowledge and representations should be built in prior to learning, in order to bootstrap the rest.

Instead of trying to build systems that learn everything from correlations between pixels and actions, we as a community need to learn how to build systems that use a core understanding of physical objects to learn about the world. Much of what we have called common sense is learned, like the idea that wallets hold money or cheese can be grated, but almost all of it starts with a firm sense of time, space, and causality. Underlying all of that may be innate machinery for representing abstraction, compositionality, and the properties of individual entities like objects and people that last over some period of time (whether measured in minutes or decades). Machines need to be wired with that much from the get-go if they are to have any chance of learning the rest.*

* Ironically, one of the great contributions to innateness within deep learning comes from one of the most anti-nativist members of the field, our NYU colleague Yann LeCun, chief scientist at Facebook. In his early work, LeCun

In a recent open letter to the field of AI, the chair of UCLA's computer science program, Adnan Darwiche, called for broader training among AI researchers, writing: "We need a new generation of AI researchers who are well versed in and appreciate classical AI, machine learning, and computer science more broadly, while also being informed about AI history."

We would extend his point, and say that AI researchers must draw not only on the many contributions of computer science, often forgotten in today's enthusiasm for big data, but also on a wide range of other disciplines, too, from psychology to linguistics to neuroscience. The history and discoveries of these fields—the cognitive sciences—can tell us a lot about how biological creatures approach the complex challenges of intelligence: if artificial intelligence is to be anything like natural intelligence, we will need to learn how to build structured, hybrid systems that incorporate innate knowledge and abilities, that represent knowledge compositionally, and that keep track of enduring individuals, as people (and even small children) do.

Once AI can finally take advantage of these lessons from cognitive science, moving from a paradigm revolving around big data to a paradigm revolving around both big data and abstract causal knowledge, we will finally be in a position to tackle one of the hardest challenges of all: the trick of endowing machines with common sense.

---

argued forcefully for the introduction of an innate bias in neural networks, called convolution, which has been nearly universally adopted in computer vision. Convolution builds networks that are "translation invariant" (that is, they recognize objects in different locations) even in advance of experience.

*Sawing the wrong side of a tree limb*

# 7

# Common Sense, and the Path to Deep Understanding

The subject of today's investigation
is things that don't move by themselves.

They need to be helped along,
shoved, shifted,
taken from their place and relocated.

They don't all want to go, e.g., the bookshelf,
the cupboard, the unyielding walls, the table.

But the tablecloth on the stubborn table
—when well-seized by its hems—
manifests a willingness to travel.

And the glasses, plates,
creamer, spoons, bowl,
are fairly shaking with desire.

—WISLAWA SZYMBORSKA,
"A LITTLE GIRL TUGS AT THE TABLECLOTH"

Non Satis Scire (To know is not enough)

—MOTTO, HAMPSHIRE COLLEGE

Common sense is knowledge that is commonly held, the sort of basic knowledge that we expect ordinary people to possess, like "People don't like losing their money," "You can keep money in your wallet," "You can keep your wallet in your pocket," "Knives cut things," and "Objects don't disappear when you cover them with a blanket." We all would be surprised if we saw a dog carry an elephant, or a chair turn into a television set. The great irony of common sense—and indeed AI itself—is that it is stuff that everybody knows, yet nobody seems to know what exactly it is or how to build machines that have it.

---

People have worried about the problem of common sense since the beginning of AI. John McCarthy, the very person who coined the name "artificial intelligence," first started calling attention to it in 1959. But there has been remarkably little progress. Neither classical AI nor deep learning has made much headway. Deep learning, which lacks a direct way of incorporating abstract knowledge (like "People want to recover things they've lost") has largely ignored the problem; classical AI has tried harder, pursuing a number of approaches, but none has been particularly successful.

One approach has been to try to learn everyday knowledge by crawling (or "scraping") the web. One of the most extensive efforts, launched in 2011, is called NELL (short for Never-Ending Language Learner), led by Tom Mitchell, a professor at Carnegie Mellon and one of the pioneers in machine learning. Day after day—the project is still ongoing—NELL finds documents on the web and reads them, looking for particular linguistic patterns and making guesses about what they might mean. If it sees a phrase like "cities such as New York, Paris, and Berlin," NELL infers that New York, Paris, and Berlin are all cities, and adds that to its database. If it sees the phrase "New York Jets quarterback Kellen Clemens," it might infer the facts that Kellen Clemens plays for the New York Jets (in the present tense—NELL has no sense of time) and that Kellen Clemens is a quarterback.

As reasonable as the basic idea is, the results have been less than stellar; as an example, here are ten facts that NELL had recently learned:

- *aggressive_dogs is a mammal*
- *uzbek is a language*
- *coffee_drink_recipes is a recipe*
- *rochelle_illinois is an island*
- *shimokitazawa_station is a skyscraper*
- *steven_hawking is a person who has attended the school cambridge*
- *cotton is an agricultural_product growing in gujarat*
- *kellen_clemens plays in the league nfl*
- *n24_17* [*sic*] *and david and lord are siblings*
- *english is spoken in the city st_julians*

Some are true, some are false, some are meaningless; few are particularly useful. They aren't going to help robots manage in a kitchen, and although they might be of modest help in machine reading, they are too disjointed and spotty to solve the challenges of common sense.

Another approach to collecting commonsense knowledge, particularly trendy nowadays, is to use "crowdsourcing," which basically means asking ordinary humans for help. Perhaps the most notable project is ConceptNet, which has been ongoing at the MIT Media Lab since 1999. The project maintains a website where volunteers can enter simple commonsense facts in English. For instance, a participant might be asked to provide facts that would be relevant in understanding the story "Bob had a cold. Bob went to the doctor," and might answer with facts such as "People with colds sneeze" and "You can help a sick person with medicine." (The English sentences are then automatically converted to machine encodings through a process of pattern matching.)

Here, too, the idea seems reasonable on its face, but the results have been disappointing. One problem is that if you simply ask untrained lay people to enumerate facts, they tend to list easily found factoids like "A platypus is a mammal that lays eggs" or "Taps is a bugle call played at dusk"—rather than what computers really need: information that is obvious to humans but hard to find on the web, such as "After something is dead, it will never be alive again" or "An impermeable container with an opening on top and nowhere else will hold liquid."

A second problem is that, even when lay people can be induced to give the right kind of information, it's tough to get them to formulate it in the kind of finicky, hyper-precise way that computers require. Here, for example, is some of what ConceptNet has learned from lay people about restaurants.

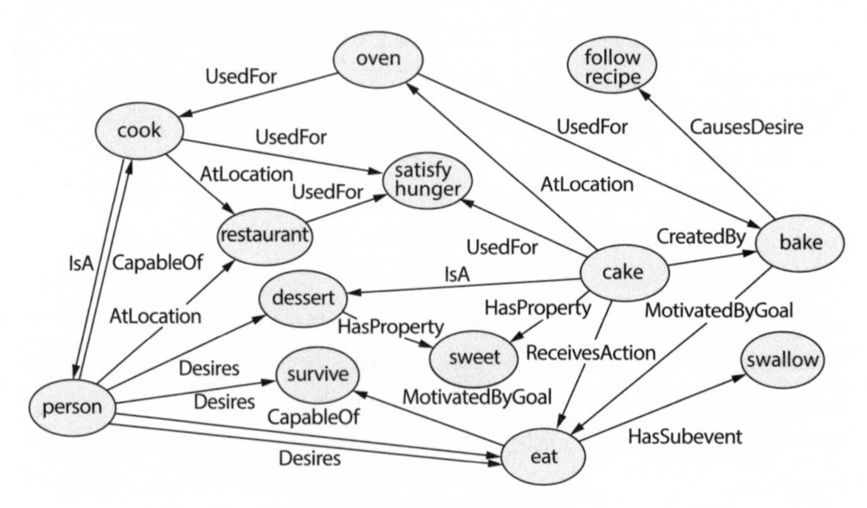


*Excerpt from ConceptNet*

To the untrained eye it seems perfectly fine. Each individual link (for example, the arrow in the top left that tells us that an oven is used for cooking) seems plausible in itself. A person can be at a location of a restaurant, and almost every person we have ever met desires survival; nobody would question the fact that we need to eat to survive.

But dive into the details and it's a mess.

Take, for example, the link that says that "person" is at location "restaurant." As Ernie's mentor Drew McDermott pointed out long ago, in a rightly famous article called "Artificial Intelligence Meets Natural Stupidity," the meaning of this sort of link is actually unclear. At any given moment, somebody in the world is at a restaurant, but many people are not. Does the link mean that if you are looking for a particular person (your mother, say) you can always find her at a restaurant? Or that at some particular restaurant (Katz's Delicatessen, say), you will always be able to find a person, 24/7? Or that any person you might care to find can always be found in a restaurant, the way that whales can always be found in the ocean? Another link tells us that a "cake UsedFor satisfy hunger." Maybe so, but beware the link that says "cook UsedFor satisfy hunger" in conjunction with "cook IsA person," which suggests that a cook might not just make a meal but become one. We're not saying that crowdsourcing couldn't ever be useful, but efforts to date have often yielded information that is confusing, incomplete, or even downright wrong.

A more recent project, also based at MIT, though run by a different team, is called VirtualHome. This project too has used crowdsourcing to collect information about procedures for simple activities like putting the groceries in the fridge and setting the table. They collected a total of 2,800 procedures (vaguely reminiscent of Schank's work on scripts, but less formally structured) for 500 tasks, involving 300 objects and 2,700 types of interactions. The basic actions were hooked into a game engine, so you can (sometimes) see an animation of the procedure in action. Once again, the results leave something to be desired. Consider for instance the crowdsourced procedure for "Exercise":

- *walk to LIVING ROOM*
- *find REMOTE CONTROL*
- *grab REMOTE CONTROL*
- *find TELEVISION*
- *switch on TELEVISION*
- *put back REMOTE CONTROL*
- *find FLOOR*
- *lie on FLOOR*
- *look at TELEVISION*

- *find ARMS_BOTH*
- *stretch ARMS_BOTH*
- *find LEGS_BOTH*
- *stretch LEGS_BOTH*
- *stand up*
- *jump*

All of that may happen in some people's exercise routine, but not in others. Some people may go to the gym, or run outside; some may jump, others may lift weights. Some steps might be skipped, others are missing; either way, a pair of stretches isn't much of a workout. Meanwhile, finding the remote control is not really an essential part of exercising; and who on earth needs to "find" their arms or legs? Something's clearly gone wrong.

Another approach is to have highly trained humans try to write it all down, in a computer-interpretable form. Many AI theorists, starting with John McCarthy and continuing today with Ernie and many of his colleagues, like Hector Levesque, Joe Halpern, and Jerry Hobbs, have tried to do just that.

Frankly, progress here too, in our home territory, has been slower than we would have hoped. The work has been both painstaking and difficult, relying on careful analysis that has thus far proven impossible to automate. Although there has been some important progress, we are nowhere close to having a thorough encoding of common sense; without it, or something close, next-level AI challenges like automated reading and home robots will continue to remain outside our grasp.

The largest effort in the field, by far, is a project known as CYC, directed by Doug Lenat, which for the last three decades has aimed at creating a massive database of human-like common sense, rendered in machine-interpretable form. It contains literally millions of carefully encoded facts on everything from terrorism to medical technology to family relationships, laboriously handcrafted by a team of people trained in AI and philosophy.

Most researchers in the field see it as a failure; far too little has been published reporting what's inside it (the project has been conducted largely in secret)—but relative to the amount of effort expended, there have been too few demonstrations of what it can do. External articles written about it have largely been critical, and very few researchers have adopted it into larger systems. We think that the goals of the project are admirable, but after three decades, CYC is still too incomplete on its own to have had an enormous impact—despite a huge amount of labor. The mystery of how to get a broad, reliable database of commonsense knowledge remains unsolved.

So what next?

We wish we had a simple and elegant answer; we don't. It is doubtful that any single or simple approach will suffice, in part because common sense itself is so diverse; no single technique is going to solve what the field has struggled with for years. Common sense is the mountain that the field needs to climb, and we have a long journey in front of us. A quick detour off the current path isn't likely to get us to the top.

That said, we do have some rough sense about where the field ought to be going. To push our mountain metaphor, if we can't get there on our own, we can at least see what the peak might look like, what equipment one might need to get there, and what sort of strategy might help.

o———o

To make progress, we need two things to get started: an inventory of what kind of knowledge a general intelligence should have, and an understanding of how this knowledge would be represented, clearly and unambiguously in a self-contained fashion, inside a machine.

We will take these on in reverse order, because finding a clear way in which to represent knowledge in a machine is a prerequisite for whatever knowledge we ultimately encode. As you might suspect by now, that task turns out to be far more subtle than it might initially seem. Some knowledge is straightforward to represent; a lot isn't.

On the easier side of the spectrum lies taxonomy, the kind of

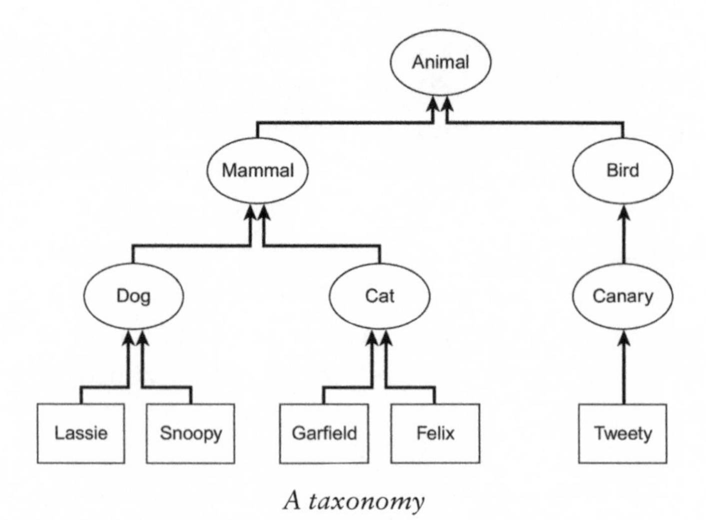


*A taxonomy*

categorization that tells us that dogs are mammals and mammals are animals, and that allows us to infer therefore that dogs are animals. If you know that Lassie is a dog and dogs are animals, then Lassie is an animal.

Online resources like Wikipedia contain vast amounts of taxonomic information: a skunk is a carnivore, a sonnet is a poem, a refrigerator is a household appliance. Another tool, WordNet—a specialized online thesaurus that is often used in AI research—lists taxonomic information for individual words: "envy" is a kind of "resentment," "milk" is both a kind of "dairy product" and a kind of "beverage," and so on. There are also specialized taxonomies, like the medical taxonomy SNOMED, with facts like "aspirin is a type of salicylate" and "jaundice is a type of clinical finding." (Online ontologies, widespread in the Semantic Web, consist in part of such knowledge.)

Similar techniques can be used to represent part/whole relationships. Told that a toe is a part of a foot and a foot is part of a body, it can infer that a toe is a part of a body. Once you know this sort of thing, a few of the puzzles we mentioned earlier start to disappear. If you see Ella Fitzgerald compared to "a vintage bottle of wine," you can take note of the fact that Fitzgerald is a person and that people are animals, and note that bottles lie in a different part of the hierarchy of things altogether (involving inanimate objects), and infer that Ella can't really be a bottle. Likewise, if a guest asks a butler robot for a drink, a robot equipped with a good taxonomy can realize that wine, beer, whiskey, or juice might qualify, but that a piece of celery, an empty glass, a clock, or a joke would not.

Alas, there's much more to common sense than taxonomy. For almost everything else, we are going to need a different approach. If the taxonomy of animal species is well defined, as a consequence of how natural selection and speciation work, many other taxonomies are not. Say we want to have a category of historical events, with individual items like "the Russian Revolution," "the Battle of Lexington," "the invention of printing," and "the Protestant Reformation." Here, the boundaries are much vaguer. Was the French Resistance part of World War II? How about the Soviet invasion of Finland in 1939? Similarly, should a category of THING that includes cars and people also include democracy, natural selection, or the belief in Santa Claus? Taxonomy is the wrong tool for the job; as Wittgenstein famously noted, it's hard to define even a simple category like "game."

And then there's the kind of knowledge we started with at the beginning of the chapter, like *knives can cut things* and *brooms can be used to clean a floor.* These facts don't seem to fit into taxonomies at all. Yet it is hard to see how a robot could take care of your home without that sort of knowledge.

Another approach has been to make the sort of diagrams we saw earlier for ConceptNet, often known as *semantic networks*. Invented in the late 1950s, semantic networks allow computers to represent a much vaster range of notions, not just which parts are parts of which wholes and which categories are inside other categories, but also all kinds of other relationships, like Albany being adjacent to the Hudson, and police officers being the sort of people who drive police cars.

But, as we already saw with ConceptNet, semantic networks' representations aren't really clear enough to solve the problem. It's much easier to draw them than to actually make them work. Suppose you want to encode facts like Ida owns an iPhone and was born in Boise, an iPhone contains a battery, and a battery produces electrical power. Pretty quickly you wind up with something like this:

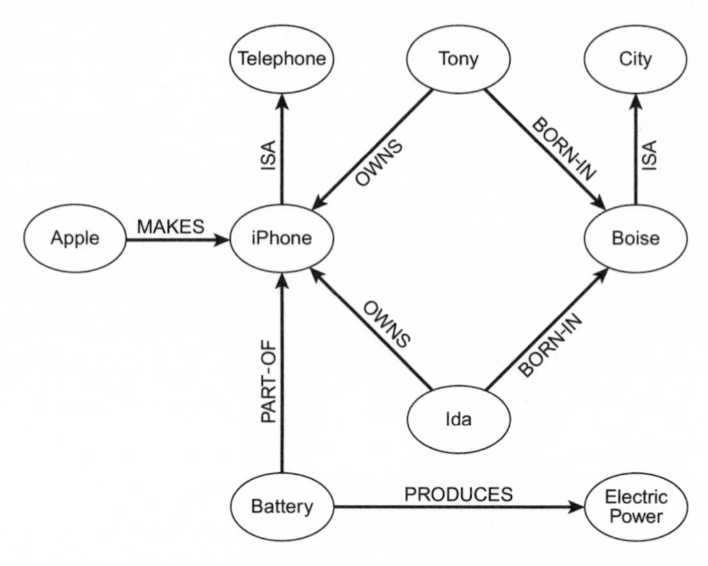


*A semantic network*

The trouble is that a lot of what you need to know to interpret the diagram is not explicit, and machines don't know how to deal with what isn't made explicit. It is obvious to you that if Tony and Ida were both born in the same Boise; it follows that if you travel to Tony's birthplace, you are in Ida's birthplace. However, if you see an iPhone that belongs to Tony, it probably doesn't also belong to Ida, yet nothing explicit in the semantic network makes that clear. Without further work, there is no way for the machine to know the difference.

Or consider the fact that Apple makes all iPhones; if you see an iPhone, you can conclude that Apple made it, which seems to follow from what's written into the network, but the network sort of makes it look like the only iPhones in the world are the two that belong to Tony and Ida, which is clearly misleading.

Every iPhone has a battery, but it also has other parts; you wouldn't conclude that every part of an iPhone is a battery, but the diagram doesn't make that clear. Digging in further, nothing in the network tells you that birthplaces are exclusive, but ownership is not exclusive. That is, if Ida was born in Boise, she couldn't also have been born in Boston but she can own both an iPhone and a TV.

Getting machines to know what's in your head as opposed to what's written on the diagram is really hard. Missing in action is the very thing that the semantic network was supposed to solve: common sense. Unless you already know about things like birth (which happens in a single location), manufacturing (one company can make more than one product), and ownership (one person can own multiple things), the formalism just doesn't help.

Things only get worse for the semantic network approach when you start thinking about time. Imagine a semantic network like this, similar to the ConceptNet stuff we discussed earlier:

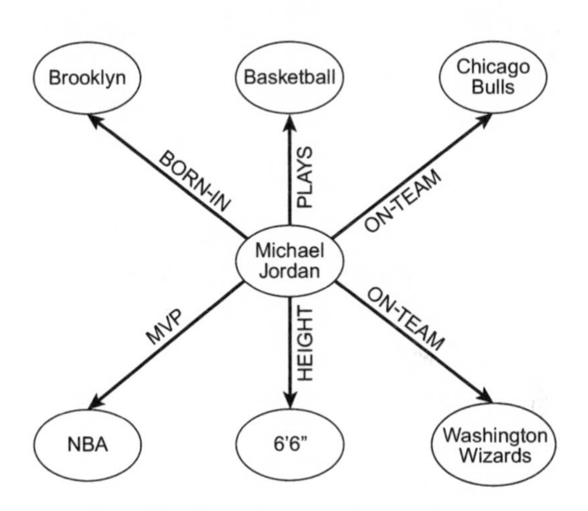


*Another semantic network*

On a superficial inspection, it all seems fine; Michael Jordan is six-foot-six, and he was born in Brooklyn, and so forth. But a deeper look reveals that a system that knew what is in this diagram and nothing more would be prone to making all sorts of nutty mistakes. It might decide that Michael Jordan was six-foot-six when he was born, or that he played for both the Wizards and the Bulls at the same time. The words "plays basketball" could refer to anything from the span of his professional career to the span from the first time he ever touched a basketball as a child to the present (assuming that he still occasionally plays casual games with his friends). If we then tell the system that Jordan played basketball from 1970 to the present, on the theory that he started when he was seven years old and continues to the present, nothing stops the system from falsely concluding that Jordan has been playing basketball 24 hours a day, 365 days a year for the last 48 years.

Until machines can deal with these sorts of things, fluidly, without humans holding their virtual hands every step of the way, they're not really going to be able to read, reason, or safely navigate the real world.

What should we do?

A first hint comes from the field of formal logic, as developed by philosophers and mathematicians. Take the assertions that "Apple makes all iPhones," "Ida owns an iPhone," and "All iPhones contain a battery." Much of the ambiguity that we saw in semantic networks can be dealt with by encoding those facts with a common technical notation:

$$\forall x \ \mathrm{IPhone}(x) \Rightarrow \mathrm{Made}(\mathrm{Apple},x)$$
$$\exists z \ \mathrm{IPhone}(z) \land \mathrm{Owns}(\mathrm{Ida},z)$$

The first statement can be read as "for every object x, if x is an iPhone, then Apple made x." The second can be read as "there exists an object z such that z is an iPhone and Ida owns z."

It takes some getting used to, and untrained people aren't comfortable with it, which makes it harder to crowdsource common sense. It's also not at all popular in AI these days; practically everyone would prefer a shortcut. But in the final analysis formal logic or something similar may be a necessary step toward representing knowledge with sufficient clarity. Unlike the semantic network link APPLE MAKES IPHONE, this formulation is unambiguous; one does not have to guess whether the first sentence means "Apple makes all iPhones," "Apple makes some iPhones" or "the only things that Apple makes are iPhones." It can only mean the first possibility.

Common sense needs to start with something like this, either formal logic or some alternative that does similar work, which is to say a way of clearly and unambiguously representing all the stuff that ordinary people know. That's a start.

Even once we nail the right way of encoding knowledge—of representing common sense in machines—we will still have challenges. One of the issues the current methods for collecting concepts and facts—hand-coding, web-mining, and crowdsourcing—all face is that they often instead wind up being a hodgepodge of facts, from "Aardvarks eat ants" to "Zyklon B is poisonous," when what we really want is for our machines to have a coherent understanding of the world.

Part of the problem is that we don't want our AI systems to go around learning every related fact individually. Rather, we want them to understand how those facts connect. We don't just want our systems to know that writers write books, that painters paint pictures, that composers compose music, and so on down the line. Instead we want those systems to see these particular facts as instances of a more general relation like "individual creates work," and to incorporate that observation itself into a larger framework that makes clear that a creator generally owns a work until he/she sells it, that works by a single person are often stylistically similar, and so on.

Doug Lenat calls these collections of knowledge "microtheories." Ultimately, there may be thousands of them; another term for them might be "knowledge frameworks." Workers in the field of knowledge representation have tried to develop such frameworks for many different aspects of real-world understanding, ranging from psychology and biology to the use of everyday objects. Although knowledge frameworks play little or no role in current big-data-centric approaches to AI, we see them as vital; a better understanding of how to build and leverage them may take us a long way.

o———o

If we could have only three knowledge frameworks, we would lean heavily on topics that were central to Kant's *Critique of Pure Reason,* which argued, on philosophical grounds, that *time, space,* and *causality* are fundamental. Putting these on computationally solid ground is vital for moving forward.

If we can't yet build them ourselves, at least we can say something about what they ought to look like.

Let's begin with time. Paraphrasing Ecclesiastes, to every event there is a time, and without an understanding of the connections between events over time, almost nothing would make sense. If a robot butler has to pour a glass of wine, then it needs to know that the cork needs to be removed before the wine can be poured, and not the other way around. A rescue robot needs to prioritize in part based on time and an understanding of how urgent various different situations are; a fire can spread in seconds, whereas it might be OK to take an hour to rescue a cat that is stuck in a tree.

Here, classical AI researchers (and philosophers) have made excellent progress, working out formal logical systems for representing situations and how they develop and change over time. A butler robot might start with the knowledge that the wine is currently in a corked bottle and the glass is currently empty and that it has a goal that within two minutes there should be wine in the glass. A so-called temporal logic can allow the robot to construct a cognitive model of these events, and then go from that cognitive model along with commonsense knowledge (for instance, if you pour from a bottle into a glass, part of the contents of the bottle will now be in the glass) to a specific plan, structured over time: uncorking the bottle at the appropriate time, pouring only afterward, and so forth, rather than the other way around.

That said, there is, as ever, serious work yet to be done. One major challenge lies in mapping external sentences onto internal representations of time. In a sentence like "Tony managed to pour the wine, though he had to use a knife to take out the cork, because he couldn't find the corkscrew," temporal logic alone isn't enough. To infer that the events mentioned occurred in backward order, you need to know something about language, as well as about time, and in particular about all the tricky ways in which sentences can describe the relations between events in time. Nobody has made substantial progress there yet. (Nor have they figured out how to integrate any of this with deep learning.)

To create a system that could figure out when Michael Jordan was and was not likely to be playing basketball or to get to a system that can reconstruct what must have happened before Almanzo returned Mr. Thompson's wallet, though, we need more than just an abstract understanding of time. It's not enough to know that events have beginnings and ends; you have to grasp particular facts about the world. When you read "Almanzo turned to Mr. Thompson and asked, 'Did you lose a pocketbook?,' " you mentally fill in a series of facts about chronology: at an earlier point in time Mr. Thompson had the wallet; at a later point in time he no longer had it; later still Almanzo found it. When you think about Michael Jordan, you use the fact that even the most avid athlete only pursues their sport in a fraction of their waking hours. To reason about how the world unfolds over time, an AI will need to integrate a complex mix of general truths (such as "a person can't perform most complex skills effectively while asleep") with specific facts, in order to figure out how those general truths apply in specific circumstances.

Likewise, no machine can yet watch a typical film and reliably sort out what is flashback and what is not. Even in films in which events mostly follow a straightforward sequence there are lots of challenges: the relations in time between one scene and the next (has a minute passed? A day? A month?) are almost always left to the viewer to figure out. This draws both on our basic understanding of how time works and on extensive detailed knowledge about what is plausible.

Once we can bring this all together, a whole world will be unlocked.

Your calendar will get a whole lot smarter, reasoning about where you need to be and when, and not just storing every event as so many isolated appointments, occasionally leaving you with far too little time to get from point A to point B; if you schedule an event in another city, the time zone will be set correctly and will not make you three hours early for a meeting; and programmers won't need to foresee this particular scenario in advance, because the AI will work out what you need from general principles. Your digital assistant will be able to tell you the list of judges currently on the Supreme Court (or any other court you choose), name the member of the Chicago Bulls roster who has played longest with the team, tell you how old Neil Armstrong was when John Glenn orbited the earth, and even tell you when you should go to bed if you have a 6:30 a.m. train to catch tomorrow and you want to get eight hours of sleep. Personalized medicine programs will be able to relate what is happening to a patient over minutes and hours to what has happened to them over a lifetime. The kind of high-end thoughtful planning that executive assistants do for CEOs will be available to everyone.

---

Machines also need a way of understanding space and the geometry of people and objects. As with time, some parts of the basic framework are well known; but there are many fundamentals that still haven't been captured. On the positive side of the ledger, Euclidean space is well understood, and we know how to do all kinds of geometric calculations. Modern specialists in computer graphics use geometry to compute how patterns of light fall onto objects in complex rooms; the results are so realistic that moviemakers routinely use their techniques to build convincing images of events that never happened in the real world.

But there's much more to understanding how the world works than just knowing how to create realistic images. Take the things you need to know about the shapes of the two ordinary objects depicted

*Ordinary objects that pose challenges for AI*

here (above), a hand grater and a mesh bag of vegetables, and what their shapes entail.

Both of these everyday objects have a fairly complex shape (much more so than a basic geometric solid like a sphere or a cube), and their shape in space matters greatly for what you can do with them. The grater is a truncated pyramid to make it stable and the handle is there so that you can hold it fixed while you grate. The pattern of holes, connecting the outside to the empty interior, allows the cheese (say) to be carved into narrow strips and fall inside the grater. Finally, on different sides the holes have their own detailed shape, to enable effective grating; for example, the circular "cheddar cheese" holes on the side facing forward in the image each have a small circular "lip" with a sharp blade so that a small strip of cheese gets caught and sliced away from the main stick. It's a pretty clever design when you think about it, and its shape in space is what governs its function.

Standard programs for graphics or computer-aided design can represent the shape, use it in a video game, calculate its volume, and even figure out which holes are in contact with some particular piece of cheese held at a particular place, but they can't reason about the functionality of the shape. We don't yet have a system that could look at a grater and understand what it is for, or how one might actually use it in order to grate mozzarella for the purpose of making a pizza.

And in some ways, the string bag is even more problematic, at least for the current state of AI. The grater at least *has* a fixed shape; you can move the grater around, but you can't bend it or fold it, so the elements of the grater remain in a constant relationship with one another. In contrast, the shape of the string bag isn't constant; it bends around the objects within in it and accommodates itself to the surfaces on which it is placed. In this sense, a mesh bag is not one specific shape, it's an infinite collection of possible shapes; all that is fixed is the lengths of the strings and the way they're connected together. From that bit of information, the AI needs to realize that you can put a cucumber and peppers into the bag, and they will stay in the bag; that you can put a pea into the bag, but it will not stay in the bag; and that you cannot put a large watermelon into the bag. Even this basic problem has never been solved. Once it is, robots will be able to work safely and effectively in busy, complex, and open-ended environments, ranging from kitchens and grocery stores to city streets and construction sites, vastly expanding their utility.

o———o

Causality, broadly interpreted, includes any kind of knowledge about how the world changes over time.* It can range from the very general—Newton's theory of gravity, Darwin's theory of evolution—to the very specific—pressing the power button on the TV remote turns the TV on and off; if a U.S. citizen does not file their

* The term "causality" is also used more narrowly to mean specifically relations of the form "A causes B," such as flipping a switch [A] completes a circuit that causes electricity to flow to the lightbulb [B]. A statement like "An object inside a closed container cannot come out" is part of a causal theory in the broad sense in which we are using the term, since it constrains what can happen over time, but not in the narrow sense, since it is not cast in terms of one event causing another. An adequate general AI needs to be able to cope with causality in both the broad and narrow senses.

annual income tax return by April 15 of the following year, the person risks penalties. The changes that occur can involve physical objects, people's minds, social organization, or virtually anything else that changes over time.

We use causality in understanding people and other creatures (psychologists call this "intuitive psychology"). We use it when we understand tools like hammers and drills, and more generally when we try to understand artifacts, human-made objects such as toasters, cars, and television sets. When we understand computers, we often treat them as artifacts that have a psychology (the machine "wants" me to enter my password; if I type in my password, the machine will recognize it and allow me to state my next request). We also use causal reasoning to understand social institutions (if you want to borrow a book, you go to the library; if you want to pass a law, you need to go through Congress), commerce (if you want a Big Mac, you will have to pay for it), contracts (if a contractor abandons a project halfway, you can sue for breach of contract), and language (if two people don't speak the same language, they can use an interpreter), and to interpret counterfactuals (if the subway workers went on strike, what would be the best way to get to work?). A very large fraction of commonsense reasoning rests on one form of causality or another, almost always revolving around time, and frequently involving space.

One hallmark of our ability to think causally is its versatility. We can, for instance, use any particular fact about causal relations in many different ways. If we understand the relation between remote control and TV, we can immediately make predictions and plans, and find explanations. We can predict that if we push the power button on the remote, the TV will turn on. We can decide that if we want to turn on the TV, we can accomplish that by pushing the right remote button. If we observe that the TV suddenly turns on, we can infer that someone else in the room probably pushed the button. The fluidity with which humans can do this, without any sort of formal training whatsoever, is startling. The ability of AIs to do the same will be revolutionary; emergency robots, for example, will be able to fix bridges and mend broken limbs with whatever materials are at hand, because they will understand materials and engineering.

o———o

Of particular importance will be the ability to fluidly combine domains of causal understanding, which is something humans do quite naturally.

Take, for example, a scene from the TV show *LA Law* that Steven Pinker described in his book *How the Mind Works.* The ruthless attorney Rosalind Shays walks into an elevator shaft; we soon hear her scream. In rapid succession, we, the audience, use physical knowledge to reason that she will fall to the bottom; we use biological knowledge to reason that such a fall would probably kill her; and we use psychological knowledge to reason that, since she did not appear suicidal, her entry into the elevator-less elevator shaft was probably a miscalculation, based on the assumption—generally true, but tragically false in this instance—that an elevator car would be present when the elevator doors opened.

Elder-care robots, for instance, will get much better when they too can fluidly understand interactions between multiple domains, in unpredictable ways. A robotic assistant, for example, has to anticipate Grandpa's psychology (How will he respond to the robot? Will he squirm? Shake the robot? Run away?) and interpret him as a complex, dynamic physical object. If the goal is getting Grandpa into bed, it's not enough to put Grandpa's center of mass on the bed if his head flops over and bangs against the railing on the side of the bed; robots that can fluidly reason about both psychology and physics will be a major advance over what we have now.

And the same fluidity will play a major role in developing machines capable of rich understanding. An AI reading the Almanzo story, for example, has to understand that Mr. Thompson initially didn't know that he had lost his wallet, and that after hearing Almanzo ask him and feeling his pocket, he does know it. That is, the AI has to infer Mr. Thompson's mental state and how that state changed over time. The reader similarly has to understand that Mr. Thompson will be upset at losing the wallet and relieved at getting it back and seeing that all the money is there, and that Almanzo will be insulted at being suspected of being a thief—again a matter of understanding the psychology of human beings, and how they feel about things like money and social interaction. With a rich enough system for causality in place, all this will come naturally.

Integrating time and causality will be pivotal when it comes to planning actions, which often requires dealing with an open-ended world and plans that are vague or underspecified; recipes, for example, often leave out steps that are too obvious to mention. Today, robots are literalists; if it's not specified, they don't do it (at least using currently available technology). In order to achieve maximum utility, they need to be as flexible as we are.

Take the process of cooking scrambled eggs. It's easy to find a recipe, but almost all recipes leave a bunch of things unsaid. One website gives the following recipe:

> BEAT eggs, milk, salt and pepper in medium bowl until blended.
> HEAT butter in large nonstick skillet over medium heat until hot.
> POUR IN egg mixture. . . .
> CONTINUE cooking—pulling, lifting and folding eggs—until thickened and no visible liquid egg remains. Do not stir constantly.

Relying on the intelligence of their readers, the authors of the recipe have skipped a lot of obvious steps that are actually involved: getting the eggs, milk, and butter out of the fridge; cracking the egg into the bowl, opening the milk carton, pouring the milk into the bowl, shaking the salt and pepper into the bowl, putting the unused eggs, milk, and butter back into the fridge, cutting the butter into the pan, turning on the flame, turning off the flame, and transferring the eggs onto a plate. If the bowl or the pan is dirty, they have to be washed before being used. If you're out of pepper, your scrambled eggs will be a little less tasty, but if you're out of eggs, you're out of luck.

More broadly, robots will need to match humans in their overall cognitive flexibility. We make plans, and then we adjust them on the fly when circumstances aren't quite what we envisioned. We can also make lots of educated guesses about scenarios we might never have experienced: What might happen if we check our luggage without remembering to close the zipper? If we try to carry a cup of coffee that is too full on a moving train? Getting robots and empowered digital assistants to make plans that are equally adaptive will be a major advance.

---

One obvious—but ultimately disappointing—way to approach causality might be through computer simulation. If we want to know whether a dog can carry an elephant, we might consider running a "physics engine" like the ones that are common in modern video games.

In certain circumstances, simulation can be an effective way of approaching causality. A physics engine can effectively generate a "mental movie" of what is happening; determining in complete detail how everything in a scenario moves and changes over time. For example, the physics engine used in a video game like *Grand Theft Auto* simulates interactions between cars, people, and every other entity in the game world. The simulator begins with a complete specification of some situation at the starting time: the exact shape, weight, material, and so forth of every object in the game. The program then uses precise physics to predict how every object will move and change from one millisecond to the next, updating things as a function of the decisions the game player makes. In essence, this is a form of causal reasoning; given this array of objects at time $t$, this is what the world must look like at time $t+1$.

Scientists and engineers often use simulation to model extremely complex situations such as the evolution of galaxies, the flow of blood cells, and the aerodynamics of helicopters.

In some cases, simulation can work for AI, too. Imagine that you are designing a robot that is picking objects off of a conveyor belt and placing those objects into boxes. The robot has to anticipate what will happen in various situations, predicting for example that in a certain position the object that it has just lifted is likely to topple over. For problems like this, a simulation can help.

However, for a number of reasons, in much of the causal reasoning that AIs must do, simulations just won't work.

You can't simply simulate everything from the level of individual atoms because the amount of computer time and memory required would be too vast. Instead, physics engines rely on shortcuts to approximate complex objects rather than deriving every detail at the atomic level. Those shortcuts turn out to be laborious to build, and for many mundane physical interactions they don't currently exist; consequently no existing physics engine is remotely complete, or likely to be complete anytime in the near term. We will need to supplement exact physical simulation with other methods.

Daily life is full of all kinds of objects that no one has bothered to build into physics engine models. Think about the operation of cutting things. A well-stocked home, with bathroom, kitchen, and tool cabinet, has a dozen or more different tools whose sole function is to cut or chop something: nail clippers, razors, knives, graters, blenders, grinders, scissors, saws, chisels, lawnmowers, and so on.

To be sure, websites called "asset stores" sell 3-D models that can be downloaded and plugged into standard physics engines, but simulators that capture sufficient detail exist for only a very small part of what a robot would encounter in everyday life. It would probably be hard to find an adequately detailed 3-D model of, say, a Cuisinart blender. If you did find it, it is very unlikely that your physics engine could predict what it will do if you try to use it to blend yogurt, banana, and milk into a smoothie; and it is an altogether safe bet that your physics engine wouldn't be able to predict what would happen if you tried to use the blender on a brick. An asset store might sell you a model of an elephant and a dog, but your physics engine probably can't predict correctly what will happen if you put the elephant on the dog's back.

Returning to scrambled eggs, it is pretty clear that people, in general, don't simulate what is going on in complex chemical and physical details and that kitchen robots shouldn't have to either. Plenty of people in this world can cook a fine plate of scrambled eggs, but probably only a tiny minority could explain the physical chemistry of how an egg cooks. Yet somehow we get by, and can do even larger tasks (like assembling a brunch) efficiently, without an exact understanding of the physics involved.*

Simulation has particularly fallen short when it comes to robots. Robots are extremely complicated mechanisms, with multiple moving parts that interact with one another—and with the outside world—in many different ways. The more interactions there are, the harder it is to get things right. For example, the Boston Dynamics dog-shaped robot SpotMini has seventeen different joints, which have several different modes of rotating and exerting force.

Moreover, what robots do typically is a function of what they perceive, in part because most motions are governed by feedback. How much force would be applied to some joint depends on what some sensor is telling the robot's arm. To use simulation to predict the consequences of a robot's action would thus entail simulating the robot's perceptions and how those would change over time. Suppose we're using a simulator to test whether a rescue robot can be counted on to carry injured people to safety in a fire. The simulator needs to

* Tellingly, even though we don't know the exact physics, we are not entirely ignorant either; scrambling an egg does not create a sort of physics-free zone where absolutely anything could happen. We know that lifting the egg out with a fork when it's cooked will be very different from trying to lift it out with a fork when it's raw, and we would be astounded if a partly cooked egg were to suddenly transform itself into an elephant. A good AI should perhaps emulate humans in this regard, with a general and flexible and largely valid understanding, even when not every detail is known.

know, not only whether the robot could physically do this, but also whether it will be able to find its way when the building is filled with smoke and the electricity is out. For now, anyway, this is far outside the scope of what can be done. More broadly, what happens to a robot when you let it loose in the real world is often completely at odds with what happens in simulation. This happens so frequently that roboticists have coined a term for it: the "reality gap."

Things become much worse for a pure simulation approach when we try to build physics engines to reason about people's minds. Suppose you want to figure out what happened when Mr. Thompson touched his pocket. In principle you might imagine simulating every molecule in his being, emulating the feedback Thompson would receive through his fingers if there was (or was not) a wallet, subsequently simulating the patterns of neural firing that would ensue, ultimately sending messages to his prefrontal cortex, culminating in a motor control program that would lead him to move his lips and tongue in a fashion that exclaimed "Yes I have! Fifteen hundred dollars in it, too."

It's a nice fantasy, but in practice that's just not possible. The amount of computing power that would be required to model Thompson in that level of detail is simply too vast. At least for now, in the early twenty-first century, we have no idea how to simulate the human brain in such detail, and because the simulator would need to calculate the interactions between so many molecules, capturing even a second in the life of Thompson's nervous system might take decades of computer time. We need systems that abstract from the exact physics, in order to capture psychology.

---

The final part of common sense is the ability to *reason.*

Think about a famous scene in the film *The Godfather*. Jack Woltz wakes up and sees a severed head of his favorite horse at the foot of the bed. Immediately he gets what Tom Hagen is telling him: If Hagen's crew can get to Woltz's horse, Hagen's crew can just as easily get to Woltz.

When we see the horse's head on Jack Woltz's bed for the first time, we don't scour our memory banks for similar examples. We (and Woltz) reason about what might happen next, drawing on a large fund of general knowledge, about how the world works, synthesizing what we know about people, objects, time, physics, economics, tools, and so forth and so on.

One of the benefits of something like formal logic is that it allows much of what we need to know to be figured out straightforwardly. It's an easy matter to infer that if a finger is part of a hand, and a hand is a part of a body, then a finger is a part of a body; the latter fact doesn't need to be crowdsourced if you know the former two and the rules of logic.

To take another example, reasoning about Rosalind Shays's death could readily be carried out by a logical inference engine that was armed with the following facts:

- An object in an empty elevator shaft is unsupported.
- The bottom of an elevator shaft is a hard surface.
- An unsupported object will fall, gaining speed rapidly.
- An object falling in an empty elevator shaft will soon collide with the bottom.
- A person is an object.
- A person who is moving quickly and collides with a hard surface is likely to be killed or badly injured.
- A person, Rosalind Shays, stepped into an empty elevator shaft.

An inference engine can then infer that Rosalind Shays was likely killed or badly injured, without having to build a full, computationally costly model of every molecule in her body.* When formal logic works, it can be a massive shortcut.

* Strictly speaking, you would need a more complex knowledge base than we supply here. For example, the first statement in the preceding list does not hold for objects that are already at the bottom of an elevator shaft, and

Logic faces its own challenges, though. For one thing, not every inference that a machine could derive is useful or relevant. Given the rule "The mother of a dog is a dog" and the fact "Lassie is a dog," a naïve system could get stuck chasing down useless and irrelevant consequences, like "Lassie's mother was a dog," "Lassie's mother's mother was a dog," and "Lassie's mother's mother's mother was a dog"—all of which are true, but unlikely to have real-world consequences. Likewise, a naïve inference engine that was trying to understand why Mr. Thompson was slapping his pocket might go down blind alleys, inferring that Mr. Thompson's pocket might be in his pants; that he probably bought the pants at a clothing store; that the clothing store had an owner at the time that Mr. Thompson bought his pants; that the owner of the clothing store probably had breakfast on the day that Mr. Thompson bought his pants; and so on, none of which would matter much for the question at hand. Cognitive scientists sometimes call this sort of challenge "the frame problem," and it is a central focus of the field of automated reasoning. Although the problem has not fully been solved, significant progress has been made.

Perhaps an even greater challenge is this: the goal of formal logical systems is to make everything precise; but in the real world, a great deal of what we need to cope with is vague. The problem of deciding whether the 1939 Soviet invasion of Finland was part of World War II is no easier in a logical notation than it was in a taxonomy. More broadly, formal logic of the sort we have been talking about does only one thing well: it allows us to take knowledge of which we are certain and apply rules that are always valid to deduce new knowledge of which we are also certain. If we are entirely sure that Ida owns an iPhone, and we are sure that Apple makes abso-

---

would require a more precise statement. It's also a good example of why it is challenging to specify exactly the correct knowledge in a way that is both flexible and sufficient.

lutely all iPhones, then we can be sure that Ida owns something made by Apple. But what in life is absolutely certain? As Bertrand Russell once wrote, "All human knowledge is uncertain, inexact, and partial."

Yet somehow we humans manage.

When machines can finally do the same, representing and reasoning about that sort of knowledge—uncertain, inexact, and partial—with the fluidity of human beings, the age of flexible and powerful, broad AI will finally be in sight.

Getting reasoning right, finding the right way to represent knowledge, and focusing on the right domains (like time, space, and knowledge) are all part of the solution, and part of what can help get us to rich cognitive models and deep understanding—the things we need most to change the paradigm.

To get there, we will need something else too: a fundamental rethinking of how learning works. We need to invent a new kind of learning that leverages existing knowledge, rather than one that obstinately starts over from square one in every domain it confronts.

In current machine-learning work, the goal is often the opposite: researchers and engineers focus on some specific, narrow task, trying to bootstrap it from scratch. The fantasy is that some magical system (which doesn't currently exist in any form) will eventually learn all it needs to know just by watching YouTube videos, with no prior knowledge. But we see no evidence that this will ever work, or even that the field is making progress in this direction.

At best, it's an empty promise: current AI video understanding is far too coarse and imprecise. A surveillance system, for example, may be able to identify the difference between a video feed in which a person is walking and one in which a person is running, but no system can reliably discern more subtle differences, like the difference between unlocking a bicycle and stealing a bicycle. Even in the best case, all that current systems can really do is *label* videos, generally poorly, with lots of errors of the sort we saw earlier. No extant system could watch *Spartacus* and have the slightest idea what's going on, or infer from a video of the Almanzo story that humans like money or don't like losing their wallets, or absorb all the information in Wikipedia and the larger web about wallets and human beings to improve its prospects. Tagging a video is not the same thing as understanding what's going on or the same as accumulating knowledge over time about how the world works.

In our minds, anyway, the idea that a so-called unsupervised video system could watch *Romeo and Juliet* and tell us about love and irony and human relationships is preposterous, certainly light-years from where we are now. So far the best we can ask are narrow technical questions like "What frame might come next in this video?" If we were to ask "What would have happened if Romeo had never met Juliet?" current systems, which lack any kind of knowledge about human relationships, would have no basis whatsoever to answer. It would be like asking a flounder to shoot a basketball.

On the other hand, we don't want to toss the baby with the bathwater; a more sophisticated, knowledge-based approach to learning is clearly critical if we are to progress. As Lenat's experience with CYC has shown, it's probably not realistic to encode by hand everything that machines need to know. Machines are going to need to learn lots of things on their own. We might want to hand-code the fact that sharp hard blades can cut soft material, but then an AI should be able to build on that knowledge and learn how knives, cheese graters, lawn mowers, and blenders work, without having each of these mechanisms coded by hand.

Historically, AI has careened between two poles: hand-wiring and machine-learning. Learning how a lawn mower works by analogy to how a knife works is very different from refining a system that categorizes dog breeds by accumulating more labeled photos, and too much research has been on the latter to the exclusion of the former. Labeling pictures of knives is just a matter of learning about common patterns of pixels; understanding what a knife *does* requires a much deeper knowledge of form and function, and how they interrelate. Comprehending the uses (and dangers) of a knife is not about accumulating lots of pictures, but about understanding—and learning—causal relationships. A digital assistant that could plan a wedding shouldn't just know that people *typically* need to bring knives and cakes, it should know *why*—that the knives are there to cut the cake. If the cake were replaced by special wedding milkshakes, the knife might not be necessary, no matter how highly correlated knives and weddings have been in prior data. Instead, a robust digital assistant ought to have enough understanding about milkshakes to recognize that the knife can stay home and an extra batch of straws might be needed. To get there, we need to take learning to a new level.

Ultimately, the lesson from studying the human mind is that we should be looking for a compromise: not blank slates that must learn absolutely everything from scratch, or systems that are fully specified in advance for any conceivable contingency, but carefully structured hybrid models with strong, innate foundations that allow systems to learn new things at a conceptual and causal level; systems that can learn theories, and not just isolated facts.

The "core" systems that Spelke has emphasized, such as systems for tracking individual people, places, and objects—standard in classical AI, but almost entirely eschewed in machine learning—seem like a good place to start.

In short, our recipe for achieving common sense, and ultimately general intelligence, is this: Start by developing systems that can represent the core frameworks of human knowledge: time, space, causality, basic knowledge of physical objects and their interactions, basic knowledge of humans and *their* interactions. Embed these in an architecture that can be freely extended to every kind of knowledge, keeping always in mind the central tenets of abstraction, compositionality, and tracking of individuals. Develop powerful reasoning techniques that can deal with knowledge that is complex, uncertain, and incomplete and that can freely work both top-down and bottom-up. Connect these to perception, manipulation, and language. Use these to build rich cognitive models of the world. Then finally the keystone: construct a kind of human-inspired learning system that uses all the knowledge and cognitive abilities that the AI has; that incorporates what it learns into its prior knowledge; and that, like a child, voraciously learns from every possible source of information: interacting with the world, interacting with people, reading, watching videos, even being explicitly taught. Put all that together, and that's how you get to deep understanding.

It's a tall order, but it's what has to be done.

# 8

# Trust

Gods always behave like the people who make them.

—ZORA NEALE HURSTON, *TELL MY HORSE*

It is not nice to throw people.

—ANNA'S ADVICE TO THE SNOW GIANT MARSHMALLOW IN JENNIFER LEE'S 2013 DISNEY FILM *FROZEN*

As we've seen, machines with common sense that actually understand what's going on are far more likely to be reliable, and produce sensible results, than those that rely on statistics alone. But there are a few other ingredients we will need to think through first.

Trustworthy AI has to start with good engineering practices, mandated by laws and industry standards, both of which are currently largely absent. Too much of AI thus far has consisted of short-term solutions, code that gets a system to work immediately, without a critical layer of engineering guarantees that are often taken for granted in other fields. The kinds of stress tests that are standard in the development of an automobile (such as crash tests and climate challenges), for example, are rarely seen in AI. AI could learn a lot from how other engineers do business.

For example, in safety-critical situations, good engineers always design structures and devices to be stronger than the minimum that their calculations suggest. If engineers expect an elevator to never carry more than half a ton, they make sure that it can actually carry five tons. A software engineer building a website that anticipates 10 million visitors per day tries to make sure its server can handle 50 million, just in case there is a sudden burst of publicity. Failing to build in adequate margins often risks disaster; famously, the O-rings in the *Challenger* space shuttle worked in warm weather but failed in a cold-weather launch, and the results were catastrophic. If we estimate that a driverless car's pedestrian detector would be good enough if it were 99.9999 percent correct, we should be adding a decimal place and aiming for 99.99999 percent correct.

For now, the field of AI has not been able to design machine-learning systems that can do that. They can't even devise procedures for making guarantees that given systems work within a certain tolerance, the way an auto part or airplane manufacturer would be required to do. (Imagine a car engine manufacturer saying that their engine worked 95 percent of the time, without saying anything about the temperatures in which it could be safely operated.) The assumption in AI has generally been that if it works often enough to be useful, then that's good enough, but that casual attitude is not appropriate when the stakes are high. It's fine if autotagging people in photos turns out be only 90 percent reliable—if it is just about personal photos that people are posting to Instagram—but it better be much more reliable when the police start using it to find suspects in surveillance photos. Google Search may not need stress testing, but driverless cars certainly do.

Good engineers also design for failure. They realize that they can't anticipate in detail all the different ways that things can go wrong, so they include backup systems that can be called on when the unexpected happens. Bicycles have both front brakes and rear brakes partly to provide redundancy; if one brake fails, the second can still stop the bike. The space shuttle had five identical computers on board, to run diagnostics on one another and to be backups in case of failure; ordinarily, four were running and the fifth was on standby, but as long as any one of the five was still running, the shuttle could be operated. Similarly, driverless car systems shouldn't just use cameras, they should use LIDAR (a device that uses lasers to measure distance) as well, for partial redundancy. Elon Musk claimed for years that his Autopilot system wouldn't need LIDAR;

from an engineering standpoint, this seems both risky and surprising, given the limitations on current machine-vision systems. (Most major competitors do use it.)

And good engineers always incorporate fail-safes—last-ditch ways of preventing complete disaster when things go seriously wrong—in anything that is mission-critical. San Francisco's cable cars have three levels of brakes. There are the basic wheel brakes, which grab the wheels; when those don't work, there are the track brakes, big wooden blocks that push the tracks together to stop the car; and when those don't work there is the emergency brake, a massive steel rod that is dropped and jams against the rails. When the emergency brake is dropped, they have to use a torch to get the train car free again; but that's better than not stopping the car.

Good engineers also know that there is a time and a place for everything; experimentation with radically innovative designs can be a game changer when mapping out a new product, but safety-critical applications should typically rely on older techniques that have been more thoroughly tested. An AI system governing the power grid would not be the place to try out some hotshot graduate student's latest algorithm for the first time.

The long-term risk in neglecting safety precautions can be serious. In many critical aspects of the cyberworld, for example, infrastructure has been badly inadequate for decades, leaving it extremely vulnerable to both accidental failures and malicious cyberattacks.* The internet of things, ranging from appliances to cars that are connected to the web, is notoriously insecure; in one famous incident, "white hat hackers" were able to take control of a journalist's jeep as it rode down the highway. Another huge vulnerability is GPS. Computer-operated devices of all kinds rely on it, not only to power automated driving directions, but also to provide location and timing for everything from telecommunications and business

* Accidents and malicious attacks are different, but they have elements in common. A door that doesn't shut properly can be blown open in a storm or pulled open by a burglar. Analogous things are true in cyberspace.

to aircraft and drones. Yet it is fairly easy to block or spoof, and the consequences could potentially be catastrophic. We also know that the Russian government has hacked into the United States power grid; nuclear power plants; and water, aviation, and manufacturing systems. In November 2018, America's water supply was described as "a perfect target for cybercriminals." If a film director wants to make an apocalyptic sci-fi movie set in the near future, these kinds of scenarios would be much more plausible than Skynet, and almost as scary. And before long cybercriminals will try to undermine AI, too.

The challenges don't end there. Once a new technology is deployed, it has to be maintained; and good engineers design their system in advance so that it can easily be maintained. Car engines have to be serviceable; an operating system has to come with some way of installing updates.

This is no less true for AI than for any other domain. An autonomous driving system that recognizes other vehicles needs to be seamlessly updated when new models of cars are introduced, and it should be obvious enough to a new hire how to fix what the original programmer set up, if the original programmer departs. For now, though, AI is dominated by big data and deep learning, and with that comes hard-to-interpret models that are difficult to debug and challenging to maintain.

---

If general principles of robust engineering apply as much to AI as to other domains, there are also a number of specialized engineering techniques that can and should be drawn from software engineering.

Experienced software engineers, for example, routinely use modular design. When software engineers develop systems to solve a large problem, they divide the problem into its component parts, and they build a separate subsystem for each of those parts. They know what each subsystem is supposed to do, so that each can be written and tested separately, and they know how they are supposed to interact, so that those connections can be checked to make sure they are working. For instance, a web search engine, at the top level, has a crawler, which collects documents from the web; an indexer that indexes the documents by their keywords; a retriever, which uses the index to find the answer to a user query; a user interface, which handles the details of communication with the user, and so on. Each of these, in turn, is built up from smaller subsystems.

The sort of end-to-end machine learning that Google Translate has made popular deliberately tries to flout this, achieving short-term gains. But this strategy comes at a cost. Important problems—like how to represent the meaning of a sentence in a computer—are deferred to another day, but not really resolved. This in turn raises the risk that it might be difficult or impossible to integrate current systems together with whatever might be needed in the future. As Léon Bottou, research lead at Facebook AI Research, has put it, "The problem of [combining] traditional software [engineering] and machine learning remains wide open."

Good engineering also requires good metrics—ways of evaluating progress so that engineers know their efforts are really making progress. The best-known metric of general intelligence by far is the Turing test, which asks whether a machine could fool a panel of judges into thinking it was human. Unfortunately, although well known, the test is not particularly useful. Although the Turing test ostensibly addresses the real world, in an open-ended way, with common sense as a potentially critical component, the reality is that it can easily be gamed. As has been clear for decades, since Eliza in 1965, ordinary people are easy to fool using a variety of cheap tricks that have nothing to do with intelligence, such as avoiding questions by appearing to be paranoid, young, or from a foreign country with limited facility for the local language. (One recent prizewinning competitor, a program named Eugene Goostman, combined all three, pretending to be a bratty thirteen-year-old from Odessa.) The goal of AI shouldn't be to fool humans; it should be to understand and act in the world in ways that are useful, powerful, and robust. The Turing test just doesn't get at that. We need something better.

For this reason, we and many of our colleagues at places such as the Allen Institute for Artificial Intelligence have been busy in recent years proposing alternatives to the Turing test, involving a wide array of challenges, ranging from language comprehension, inferring physical and mental states, and understanding YouTube videos, elementary science, and robotic abilities. Systems that learn some video games and then transfer those skills to other games might be another step. Even more impressive would be a robot scientist that could read descriptions of simple experiments in "100 Science Experiments for Kids," carry them out, understand what they prove, and understand what would happen instead if you did them a little differently. No matter what, the key goal should be to push toward machines that can reason flexibly, generalizing what they have learned to new situations, in robust ways. Without better metrics, it would be hard for the quest for genuine intelligence to succeed.

Finally AI scientists must actively do their best to stay far away from building systems that have the potential to spiral out of control. For example, because consequences may be hard to anticipate, research in creating robots that could design and build other robots should be done only with extreme care, and under close supervision. As we've often seen with invasive natural creatures, if a creature can reproduce itself and there's nothing to stop it, then its population will grow exponentially. Opening the door to robots that can alter and improve themselves in unknown ways opens us to unknown danger.

Likewise, at least at present, we have no good way of projecting what full self-awareness for robots might lead to.* AI, like any

* If we *were* to build deeply self-aware robots, capable of introspection, self-improvement, and goal setting, one safe bet is that some important and challenging ethical questions would arise. While we have no hesitation in deleting a social-networking app that has outlived its usefulness, it's easy to see that tougher questions might arise with respect to robots that were genuinely sentient, possessing, say, as much self-awareness as a nonhuman primate. Would it be OK to damage them? To strip them for parts? To permanently turn them off? If they reached human-level intelligence, would it be necessary to grant them civil and property rights? Would they be subject

technology, is subject to the risk of unintended consequences, quite possibly more so, and the wider we open Pandora's box, the more risk we assume. We see few risks in the current regime, but fewer reasons to tempt fate by blithely assuming that anything that we might invent can be dealt with.

We are cautiously optimistic about the potential contribution to AI safety of one software engineering technique in particular, known as program verification, a set of techniques for formally verifying the correctness of programs that at least thus far is more suited to classical AI than to machine learning. Such techniques have used formal logic to validate whether a computer system works correctly, or, more modestly, that it is at least free from specific kinds of bugs. Our hope is that program verification can be used to increase the chance that a given AI component will do what it is intended to do.

Every device that plugs into a computer, such as a speaker, a microphone, or an external disk drive, requires a device driver, which is a program that runs the device and allows the computer to interact with it. Such programs are often extremely complicated pieces of code, sometimes hundreds of thousands of lines long. Because device drivers necessarily have to interact closely with central parts of the computer's operating system, bugs in the driver code used to be a major problem. (The problem was made even more acute by the fact that the device drivers were typically written by hardware manufacturers rather than by the software company that built the operating systems.)

For a long time, this created total chaos, and numerous system crashes, until eventually, in 2000, Microsoft imposed a set of strict rules that device drivers have to follow in their interactions with the

---

to criminal law? (Of course, we are nowhere near this point yet. When the chatbot/talking head Sophia was granted Saudi Arabian citizenship in 2017, it was a publicity stunt, not a milestone in self-aware AI; the system relied on the usual tricks of precrafted dialogue, not genuine self-awareness.)

Windows operating systems. To ensure that these rules were being followed, Microsoft also provided a tool called the Static Driver Verifier, which uses program verification techniques to reason about the driver's code, in order to ensure that the driver complies with the rules. Once that system was put in place, system crashes were significantly reduced.

Similar reasoning systems have been used to check for bugs of particular kinds in other large programs and hardware devices. The computerized control program for the Airbus airliners was verified—which is to say formally and mathematically guaranteed—to be free of bugs that could cause their enormously complex software to crash. More recently, a team of aerospace engineers and computer scientists from Carnegie Mellon and Johns Hopkins combined software verification with reasoning about physics to verify that the collision avoidance programs used in aircraft are reliable.

To be sure, program verification has limits. Verification can estimate how the plane will respond in different kinds of environments; it can't guarantee that human pilots will fly the planes according to protocol, or that sensors will work properly (which may have been a factor in two fatal accidents involving the Boeing 737 Max), or that maintenance workers will never cut corners, nor that parts suppliers will always meet their specifications.

But verifying that the software itself won't crash is a very important start, and far better than the alternative. We don't want our airplane's software to reboot midflight, and we certainly don't want our robot's code to crash while it's busy assembling a bookshelf, nor for it to suddenly mistake our daughter for an intruder.

AI researchers should be thinking hard about how to emulate the spirit of that work, and more than that, they should be thinking about how the tools of deep understanding might themselves open new approaches to having machines reason about the correctness, reliability, and robustness of software.

At the very least, as the technology advances, it might become possible to prove that the system avoids certain kinds of mistakes; for instance, that, under normal circumstances, a robot will not fall over or bump into things; or that the output of a machine translation is grammatically correct. More optimistically, the cognitive power of AI itself may be able to take us further, eventually emulating the ability of skilled software architects to envision how their software works in a wide range of environments, improving coding and debugging.

o———o

Every technique we have reviewed requires hard work, and more than that, patience. We belabor them (even if some may seem obvious) because the kind of patience we advocate here is too easily ignored in the heat of the moment; often, it's not even valued. Silicon Valley entrepreneurs often aspire to "move fast and break things"; the mantra is "Get a working product on the market before someone beats you to it; and then worry about problems later." The downside is that a product created this way often works at one scale, but needs to be wholly rewritten when the situation changes; or it works for the demo, but not the real world. This is known as "technical debt": you get a first, sometimes bug-ridden, version of the product you want; but you often have to pay later, with interest, in making the system robust, rooting out stopgaps and rebuilding foundations. That might be OK for a social media company but could prove dangerous for a domestic robot company. Shortcuts in a social-networking product might lead to user outages, bad for the company but not for humanity; shortcuts in driverless cars or domestic robots could easily be deadly.

Ultimately there is no single cure-all for good AI design, any more than there is for engineering in general. Many convergent techniques must be used and coordinated; what we have discussed here is just a start.

o———o

Deep-learning- and big-data-driven approaches pose an additional set of challenges, in part because they work very differently from traditional software engineering.

Most of the world's software, from web browsers to email clients to spreadsheets to video games, consists not of deep learning, but of classical computer programs: long, complex sets of instructions carefully crafted by humans for particular tasks. The mission of the computer programmer (or team of programmers) is to understand some task, and translate that task into instructions that a computer can understand.

Unless the program to be written is extremely simple, the programmer probably won't get it right the first time. Instead, the program will almost certainly break; a big part of the mission of a programmer is to identify "bugs"—that is, errors in the software—and to fix those bugs.

Suppose our programmer is trying to build a clone of *Angry Birds,* in which flaming tigers must be hurled into oncoming pizza trucks in order to forestall an obesity epidemic. The programmer will need to devise (or adapt) a physics engine, determining the laws of the game universe, tracking what happens to the tigers as they are launched into flight, and whether the tigers collide with the trucks. The programmer will need to build a graphics engine to make tigers and pizza trucks look pretty, and a system to track the users' commands for maneuvering the poor tigers. Each and every component will have a theory behind it (I want the tigers to do this and then

*Debugging a video game*

that when this other thing happens), and a reality of what happens when the computer actually executes the program.

On a good day, everything aligns: the machine does what the programmer wants it to do. On a bad day, the programmer leaves out a punctuation mark, or forgets to correctly set the first value of some variable, or any of ten thousand other things. And maybe you wind up with tigers that go the wrong way, or pizza trucks that suddenly appear where they shouldn't. The programmer herself may spot the bug, or the software might get released to an internal team that discovers the bug. If the bug is subtle enough, in the sense of happening only in unusual circumstances, maybe the bug won't be discovered for years.

But all debugging is fundamentally the same: it's about identifying and then localizing the gap between what a programmer *wants* their program to do, and what the program (executed by an infinitely literal-minded computer) is actually doing. The programmer wants the tiger to disappear the instant that it collides with the truck, but for some reason, 10 percent of the time the image of the tiger lingers after the collision, and it's the programmer's job to figure out why. There is no magic here; when programs work, programmers understand why they work, and what the logic is that they are following. Generally, once the underlying cause of the bug is identified, it's not hard to understand the logic of why something doesn't work. Hence, once the cause of the bug is found, it is often easy to remedy.

By contrast, a field like pharmacology can be very different. Aspirin worked for years before anybody had a clear idea of how it worked, and biological systems are so complex that it is rare for the actions of a medicine to be completely and fully understood. Side effects are the rule rather than the exception because we can't debug drugs the way we can debug computer programs. Our theories of how drugs work are mostly vague, at some level, and much of what we know comes simply from experimentation: we do a drug trial, find that more people are helped than harmed and that the harm is not too serious, and we decide that it is OK to use the drug.

One of the many worries about deep learning is that deep learning is in many ways more like pharmacology than like conventional computer programming. The AI scientists who work on deep learning understand, in broad terms, why a network trained over a corpus of examples can imitate those examples on new problems. However, the choice of network design for a particular problem is still far from an exact science; it is guided more by experimentation than theory. Once the network is trained to carry out its task, it is largely mysterious how it works. What one winds up with is a complex network of nodes whose behavior is determined by hundreds of millions of numerical parameters. Except in rare cases, the person who builds the network has little insight into what any of the individual nodes do, or why any of the parameters have their particular value. There is no clear explanation of why the system gets the right answer when it works correctly, or why it gets the wrong answer when it doesn't. If the system doesn't work, it's largely a trial-and-error process to try to fix things, either through subtle alterations of the network architecture, or by building better databases of training data. (For this reason there is a recent thrust in both machine-learning research and public policy toward "explainable AI," though no clear results yet.)

And vast storehouses of human knowledge that could be used to make systems better and more reliable are currently neglected, because it is far from clear how to integrate them into a deep learning workflow. In vision, we know a lot that is relevant about the shapes of objects and the way that images are formed. In language, we know a lot about the structure of language: phonology, syntax, semantics, and pragmatics. In robotics, we know a lot about the physics of robots and their interactions with external objects. But if we use end-to-end deep learning to build an AI program for these, all that knowledge goes out the window; there is simply no way to take advantage of it.

---

If Alexa had a well-engineered commonsense system in place, it would not start laughing out of the blue; it would recognize that people tend to laugh in response to particular circumstances, such as at jokes and in awkward moments. With common sense installed, Roomba would not smear dog poop around; it would recognize that a different solution was required; at the very least it would ask for help. Tay would recognize that large constituencies would be offended by its descent into hate speech, and the hypothetical butler robot would be careful not to break glasses on its way to pouring wine. If Google Images had a clearer idea of what the world is actually like, it would realize that there are many, many mothers who are not white. And, as we will explain later, with common sense, we'd also be significantly less likely to be all transformed into paper clips.

Indeed, a large swath of what current AI does that seems obviously foolish or inappropriate could presumably be avoided in programs that had deep understanding, rather than just deep learning. An iPhone would not autocorrect to "Happy Birthday, dead Theodore" if it had any idea what "dead" means and when one wishes a person Happy Birthday. If Alexa had any idea about what kinds of things people are likely to want to communicate, and to whom, it would double-check before sending a family conversation to a random friend. The estrus-prediction program would realize that it is not doing its job if it never predicts when the cows are in estrus.

Part of the reason we trust other people as much as we do is because we by and large think they will reach the same conclusions as we will, given the same evidence. If we want to trust our machines, we need to expect the same from them. If we are on a camping trip and both simultaneously discover that the eight-foot-tall hairy ape known as Sasquatch (aka Bigfoot) is real and that he looks hungry, I expect you to conclude with me, from what you know about primates and appetite, that such a large ape is potentially dangerous, and that we should immediately begin planning a potential escape. I don't want to have to argue with you about it, or to come up with ten thousand labeled examples of campers who did and didn't survive similar encounters, before acting.

Building robust cognitive systems has to start with building sys-

*Attacked by Sasquatch, while robot sorts through data in search of a plan*

tems with a deep understanding of the world, deeper than statistics alone can provide. Right now, that's a tiny part of the overall effort in AI, when it should really be *the* central focus of the field.

Finally, in order to be trustworthy, machines need to be imbued by their creators with ethical values. Commonsense knowledge can tell you that dropping a person out of a building would kill them; you need values to decide that that's a bad idea. The classic statement of fundamental values for robots is Isaac Asimov's "Three Laws of Robotics," introduced in 1942.

- A robot may not injure a human being or, through inaction, allow a human being to come to harm.
- A robot must obey the orders given it by human beings except where such orders would conflict with the First Law.
- A robot must protect its own existence as long as such protection does not conflict with the First or Second Laws.

For the many straightforward ethical decisions that a robot must make in everyday life, Asimov's laws are fine. When a companion robot helps someone do the shopping, it generally should not shoplift, even if its owners tell it to, because that would hurt the shopkeeper. When the robot walks someone home, it generally should not push other pedestrians out of the way, even though that might get the person it is accompanying home faster. A simple regimen of "don't lie, cheat, steal, or injure," as special cases of causing harm, covers a great many circumstances.

As University of Pittsburgh ethicist Derek Leben has pointed out, though, things start to get murkier in many other cases. What kind of harm or injury need a robot consider beyond physical injury: loss of property, of reputation, of employment, of friends? What kinds of indirect harm need the robot consider? If it spills some coffee on an icy sidewalk and someone later slips on it, has it violated the First Law? How far must the robot go in not allowing humans to be harmed by inaction? In the time that it takes you to read this sentence, four human beings will die; is it a robot's responsibility to try to prevent those deaths? A driverless car (which is, again, a wheeled robot) that pondered all of the places where it could be at any moment might never make it out of the driveway.

Then there are moral dilemmas, including many situations in which, whatever the robot does, someone will be injured, like the one that Gary introduced in *The New Yorker* in 2012, in homage to Philippa Foot's classic Trolley problems: What should a driverless car do if it encounters a school bus full of children spinning out of control hurtling toward it on a bridge? Should the car sacrifice itself and its owner to save the schoolchildren, or protect itself and its owner at all costs? Asimov's First Law doesn't really help, since human lives must be sacrificed one way or the other.

Real-life moral dilemmas are often even less clear-cut. During World War II, a student of the existentialist philosopher Jean-Paul Sartre was torn between two courses of action. The student felt that he should go join the Free French and fight in the war, but his mother was entirely emotionally dependent on him (his father had abandoned her and his brother had been killed). As Sartre put it: "No general code of ethics can tell you what you ought to do." Maybe someday in the distant future, we could build machines to worry about such things, but there are more pressing problems.

No current AI has any idea what a war is, much less what it means to fight in a war, or what a mother or country means to an individual. Still, the immediate challenge isn't the subtle stuff; it's to make sure that AIs don't do things that are *obviously* unethical. If a digital assistant wants to help a person in need who has little cash, what's to stop the AI from printing dollar bills on the color printer? If someone asks a robot to counterfeit, the robot might figure there is little harm; nobody who gets or spends the bill in the future will be hurt, because the counterfeit is undetectable, and it might decide that the world as a whole may be better off because the spending of the extra money stimulates the economy. A thousand things that seem utterly wrong to the average human may seem perfectly reasonable to a machine. Conversely, we wouldn't want the robot to get hung up on moral dilemmas that are more imaginary than real, pondering for too long whether to rescue people from a burning building because of the potential harm the occupants' great-grandchildren might someday inflict on others.

The vast majority of the time, the challenge for AIs will not be to succeed in extraordinary circumstances, solving Sophie's choice or Sartre's student's dilemma, but to find the right thing to do in ordinary circumstances, like "Could striking this hammer against this nail on this board in this room at this moment plausibly bring harm to humans? Which humans? How risky?" or "How bad would it be if I stole this medicine for Melinda, who can't afford to pay for it?"

We know how to build pattern classifiers that tell dogs from cats, and a golden retriever from a Labrador, but nobody has a clue how to build a pattern classifier to recognize "harm" or a "conflict" with a law.

Updated legal practices will of course be required, too. Any AI that interacts with humans in open-ended ways should be required, by law, to understand and respect a core set of human values. Existing prohibitions against committing theft and murder, for example, should apply to artificial intelligences—and those who design, develop, and deploy them—just as they do for people. Deeper AI will allow us to build values into machines, but those values must also be reflected in the people and companies that create and operate them, and the social structures and incentives that surround them.

Once all this is in place—values, deep understanding, good engineering practices, and a strong regulatory and enforcement framework—some of the field's biggest worries, like Nick Bostrom's widely discussed paper-clip example, start to dissolve.

The premise of Bostrom's thought experiment, which at first blush seems to have the feel of inexorable logic, is that a superintelligent robot would do everything in its power to achieve whatever goal had been set for it—in this case, to make as many paper clips as possible. The paper-clip maximizer would start by requisitioning all readily available metal to make as many paper clips as possible, and when it ran out, it would start mining all the other metal available in the universe (mastering interstellar travel as a step along the way), and eventually, when other obvious sources of metal have been used up, it would start mining the trace atoms of metal in human bodies. As Eliezer Yudkowsky put it, "The AI does not hate you, nor does it love you, but you are made out of atoms which it can use for something else." Elon Musk (who tweeted about Bostrom's book) seemed to have been influenced by this scenario when he worried that AI might be "summoning the demon."

But there's something off about the premise: it assumes that we will eventually have a form of superintelligence smart enough both to master interstellar travel and to understand human beings (who would surely resist being mined for metals), and yet possessed of so little common sense that it never comes to realize that its quest is

(a) pointless (after all, who would use all those paper clips?), and (b) in violation of even the most basic moral axioms (like Asimov's).

Whether it is even possible to build such a system—superintelligent yet entirely lacking in both common sense and basic values—seems to us unclear. Can you construct an AI with enough of a theory of the world to turn all the matter of the universe into paper clips, and yet remain utterly clueless about human values? When one thinks about the amount of common sense that will be required in order to build a superintelligence in the first place, it becomes virtually impossible to conceive of an effective and superintelligent paper-clip maximizer that would be unaware of the consequences of its actions. If a system is smart enough to contemplate enormous matter-repurposing projects, it is necessarily smart enough to infer the consequences of its intended actions, and to recognize the conflict between those potential actions and a core set of values.

And that—common sense, plus Asimov's First Law, and a fail-safe that would shut the AI down altogether in the event of a significant number of human deaths—ought to be enough to stop the paper-clip maximizer in its tracks.

Of course, people who enjoy the paper-clip tale can extend it endlessly. (What if the maximizer is spectacularly good at deceiving people? What if the machine refused to allow people to turn it off?) Yudkowsky argues that people who expect AIs to be harmless are merely anthropomorphizing; they are unconsciously reasoning that, since humans are more or less well-intentioned, or at least mostly don't want to wipe out the human race, that AIs will be the same way. The best solution in our view is neither to leave the matter to chance nor to have the machine infer all of its values directly from the world, which would be risky in a Tay-like way. Rather, some well-structured set of core ethical values should be *built in;* there should be a legal obligation that systems with broad intelligence that are powerful enough to do significant harm understand the world in a deep enough fashion to be able to understand the consequences of their actions, and to factor human well-being into the decisions they make. Once such precautions are in place, irrationally exuberant maximization with seriously deleterious consequences ought to be both illegal and difficult to implement.*

So for now let's have a moratorium on worrying about paper clips, and focus instead on imbuing our robots with enough common sense to recognize a dubious goal when they see it. (We should also be careful not to issue entirely open-ended instructions in the first place.) As we have stressed, there are other, much more pressing and immediate concerns than paper-clip maximizers that our best minds should be agonizing over, such as how to make domestic robots that can reliably infer which of its actions are and are not likely to cause harm.

On the plus side, AI is perhaps unique among technologies in having the logical potential to mitigate its own risks; knives can't reason about the consequences of their actions, but artificial intelligences may someday be able to do just that.

---

We both first learned about AI through science fiction as kids, and we constantly marvel at what has and what has not been accomplished. The amount of memory and computing power and networking technology packed into a smart watch amazes us, and even a few years ago we didn't expect speech recognition to become so ubiquitous so quickly. But true machine intelligence is much further from being achieved than either of us expected when we started thinking about AI.

Our biggest fear is not that machines will seek to obliterate us or turn us into paper clips; it's that our aspirations for AI will exceed

* Of course, there are gray areas. Should an advertising maximizer embodied with common sense and values prevent enemy nations from tampering with people's news feeds? Should a dating service app, constrained by a mandatory values system, be permitted to tamper with existing romantic relationships by offering its users endless temptation, in the form of a parade of ostensibly more attractive alternatives? Reasonable people might disagree about what ought to be permitted.

our grasp. Our current systems have nothing remotely like common sense, yet we increasingly rely on them. The real risk is not superintelligence, it is idiots savants with power, such as autonomous weapons that could target people, with no values to constrain them, or AI-driven newsfeeds that, lacking superintelligence, prioritize short-term sales without evaluating their impact on long-term values.

For now, we are in a kind of interregnum: narrow but networked intelligences with autonomy, but too little genuine intelligence to be able to reason about the consequences of that power. In time, AI will grow more sophisticated; the sooner it can be made to reason about the consequences of its actions, the better.

All of which connects very directly to the larger theme of this book. We have argued that AI is, by and large, on the wrong path, with the majority of current efforts devoted to building comparatively unintelligent machines that perform narrow tasks and rely primarily on big data rather than on what we call deep understanding. We think that is a huge mistake, for it leads to a kind of AI adolescence: machines that don't know their own strength, and don't have the wherewithal to contemplate the consequences of their own actions.

The short-term fix is to muzzle the AI that we build, making sure that it can't possibly do anything of serious consequence, and correcting each individual error that we discover. But that's not really viable in the long term, and even in the short term, we often (as we have seen) wind up with Band-Aids rather than comprehensive solutions.

The only way out of this mess is to get cracking on building machines equipped with common sense, cognitive models, and powerful tools for reasoning. Together, these can lead to deep understanding, itself a prerequisite for building machines that can reliably anticipate and evaluate the consequences of their own actions. That project itself can only get off the ground once the field shifts its focus from statistics and a heavy but shallow reliance on big data. The cure for risky AI is better AI, and the royal road to better AI is through AI that genuinely understands the world.

# Epilogue

Trustworthy AI, grounded in reasoning, commonsense values, and sound engineering practice, will be transformational when it finally arrives, whether that is a decade or a century hence.

Already, over the past two decades, we have seen major technological advances, mostly in the form of "blank slate" machine learning applied to big data sets, in applications such as speech recognition, machine translation, and image labeling. We don't expect that to stop. The state of the art in image and video labeling will continue to progress; chatbots will get better; and the ability of robots to maneuver and grasp things will continue to improve. We will see more and more clever and socially beneficial applications, such as using deep learning to track wildlife and predict aftershocks of earthquakes. And of course there will also be advances in less benign domains, such as advertising, propaganda, and fake news, along with surveillance and military applications, long before the reboot that we have called for arrives.

But eventually all that will seem like an appetizer. In hindsight, the turning point will be seen not as the 2012 rebirth of deep learning, but as the moment at which a solution to the challenges of common sense and reasoning yields deeper understanding.

What will that mean? Nobody knows for sure; nobody can pretend to predict what the future will look like in all its ramifications.

In the 1982 film *Blade Runner,* the world is filled with advanced AI-powered replicants that seem all but indistinguishable from humans—and yet at a crucial moment Rick Deckard (Harrison Ford) stops at a pay phone to place a call. In the real world, it's a lot easier to replace pay phones with cell phones than to build human-level AI, but nobody on the film crew considered that part of the chronology. In any set of predictions about fast-moving technologies, ours or anyone else's, there are bound to be some pretty glaring errors.

Still, we can at least make some educated guesses. To begin with, AI that is powered by deep understanding will be the first AI that can learn the way a child does, easily, powerfully, constantly expanding its knowledge of the world, and often requiring no more than one or two examples of any new concept or situation in order to create a valid model of it. It will also be the first to truly comprehend novels, films, newspaper stories, and videos. Robots embedded with deep understanding would be able to move safely around the world and physically manipulate all kinds of objects and substances, identifying what they are useful for, and to interact comfortably and freely with people.

Once computers can understand the world and what we're saying, the possibilities are endless. To begin with, search will get better. Many of the questions that confuse current technology—"Who is currently on the Supreme Court?," "Who was the oldest Supreme Court justice in 1980?," "What are the Horcruxes in *Harry Potter*?"—would suddenly be easy for machines. And so too will many questions we wouldn't dream of asking now: A screenwriter could ask some future search engine, "Find a short story involving the leader of one country being an agent of another, for screen adaptation." An amateur astronomer could ask, "When's the next time that Jupiter's Great Red Spot will be visible?," taking into account the local weather forecasts as well as the astronomy. You could tell your video game program that you want your avatar to be a rhinoceros wearing a tie-dyed shirt, instead of choosing from a dozen preprogrammed options. Or ask your e-reader to track every book you read, and sort how much time you spend reading literature from each continent, broken down by genre.

Digital assistants, meanwhile, will be able to do pretty much anything human assistants can do, and they will become democratized, available to all rather than just the wealthy. Want to plan a corporate retreat for a thousand employees? Your digital assistant—equipped with deep understanding—will do most of the work, from figuring out what needs to be ordered to who needs to be called and reminded, encompassing what Google Duplex hopes to do (dialing and interacting with people on the other end) but not just for a haircut or a dinner reservation with a preordained script, but for a vast, customized operation that might involve dozens of staff and subcontractors, from chefs to videographers. Your digital assistant will be both liaison and project manager, managing parts of a hundred people's calendars, and not just your own.

Computers will also become vastly easier to use. No more poring through help menus and memorizing keyboard shortcuts. If you suddenly want to italicize all foreign words, you could just ask, instead of going through the whole document yourself, word by word. Want to copy forty different recipes from forty different web pages, automatically converting all imperial measurements into metric, and scaling everything for four portions? Instead of hunting for an app that has just the right set of features, all you will have to do is ask, in English, or whatever language you prefer. Essentially anything tedious that we currently do with our computers might be done automatically, and with much less fuss. Web pages full of "Chrome annoyances" and "PowerPoint annoyances" that detail minor but important contingencies that corporate software developers forgot to anticipate, much to the aggravation of their users, will disappear. In time, that new freedom will seem as life-changing as web search, and probably more so.

The *Star Trek* holodeck will become a reality, too. Want to fly over Kilauea while it's erupting? Or to accompany Frodo to Mount Doom? Just ask. The rich virtual reality worlds imagined in the book and film *Ready Player One* will become something anyone can try.

We already know how to make graphics that good; AI powered with deep understanding will make rich, intricate human-like characters possible, too. (And for that matter, psychologically rich aliens, too, who make plausible choices given bodies and minds very different from our own.)

Meanwhile, domestic robots will finally become practical—and trustworthy—enough to work in our homes, cooking, cleaning, tidying, buying groceries, and even changing lightbulbs and washing windows. Deep understanding might also be just the ticket for making driverless cars genuinely safe, too.

Over time, the techniques that allowed machines to have an ordinary person's understanding of the world could be extended, to match the even richer understanding of human experts, extending beyond basic common sense and into the sort of technical expertise that a scientist or doctor possesses.

When this understanding is achieved, perhaps decades from now, after a great deal of hard work, machines will be able to start doing expert-level medical diagnosis, digest legal cases and documents, teach complex subjects, and so forth and so on. Of course, political problems will remain—hospitals will still have to be persuaded that the economics of better AI makes sense, and better sources of energy, even if they are invented by machines, still have to be adopted. But once AI is good enough, many technical challenges will be surmountable for the first time.

Computer programming might finally be automated too, and the power of any one individual to do something new, like build a business or invent an art form, will be vastly greater than it is now. The construction industry will change too, as robots start to be able to do the skilled work of carpenters and electricians; the time required to build a new house will be reduced, and the cost will decrease, too. Nearly anything that is dirty and dangerous, even if it requires expertise, will become automated. Rescue robots and robotic firefighters will be widespread, delivered from the factory with skills ranging from CPR to underwater rescue.

Artists, musicians, and hobbyists of all stripes will be empowered by AI-powered assistants that can radically enhance the scope of any project. Want to practice your Beatles tunes with a robot band, or conduct a live symphony orchestra, consisting of robots that play in perfect time? Play doubles tennis with your spouse against licensed replicas of the Williams sisters? No problem. Build a life-size castle made out of Legos, with robots that joust? A formation of flying drones that replicates Stonehenge for your next visit to Burning Man? AI will be there to assist with every calculation you require; robots will do most of the labor. Individuals in virtually every field will be able to do things they never could have imagined; each one will be able to serve as the creative director for a whole team of robot helpers. (People will also have more free time, with AI and robots doing much of the tedious work of everyday life.)

Here again, of course, things may not proceed in lockstep. We may well achieve expert-level deep understanding in some domains before others; particular areas of quantitative science, say, might see advances, even as AI still struggles to reach child-level abilities in others. The perfect musical helper might well come years before the perfect AI paralegal.

The ultimate, of course, is a machine that can teach itself to be an expert, in any domain. That too will come, in time.

o———o

Eventually, the pace of scientific discovery itself might vastly accelerate, once we combine the sheer computational power of machines with software that can match the flexibility and powerful intuitions of human experts.

At this point, a single advanced computer might replicate what a team of carefully trained humans can do, or even do things that humans simply can't, because we can't, for example, track in our heads the interrelations of thousands of molecules, particularly not with the kind of mathematical precision that comes naturally to machines. With such advanced forms of AI, it might finally be possible to use sophisticated causal inferencing in conjunction with vast amounts of neural data to figure out how the brain works (far too little is known now) and how to make drugs that cure mental disorders (where hardly any progress has been made in three decades). It's not unreasonable to think that AI with genuine scientific prowess could also help devise more efficient technologies for agriculture and clean energy. None of this will come soon or easily, because AI is so hard, but it will come eventually.

o———o

Which is not to say that all will be good. On the plus side, if Peter Diamandis is right, with all that automation will come abundance, and the price of many things, ranging from groceries to electricity, may come down. In the best case, humanity might well attain the vision of Oscar Wilde: "amusing itself, or enjoying cultivated leisure . . . making beautiful things, or reading beautiful things, or simply contemplating the world with admiration and delight, [with] machinery . . . doing all the necessary and unpleasant work."

Realistically, though, employment is likely to become scarcer, and debates about guaranteed basic income and income redistribution are likely to become even more pressing than they are now. Even if the economic issues can be resolved, many people may have to change how they derive a sense of self-worth, from a model that is based in large part on work, to a model that finds fulfillment through personal projects like art and creative writing, once a vast amount of nonskilled labor is automated. Of course there will be new jobs (like robot repair, which at least initially may be difficult to automate), but it seems unwise to assume that the new professions will fully take the place of older ones. The fabric of society may very well change, with increased leisure time, lower prices, and decreased drudgery but also fewer employment opportunities, and possibly greater income inequality. It's fair to expect that a cognitive revolution in AI will spark as many changes in society as the Industrial Revolution did, some positive, some not, many quite dramatic. Solving AI won't be a panacea, but we do think it will generally be positive, with all that it can do for science, medicine, the environment, and technology, provided that we are smart and cautious about the way we pursue it.

Does this mean our descendants will live in a world of abundance in which machines do almost all the hard work, and we live lives of cultivated leisure, à la Wilde and Diamandis? Or a world in which we upload copies of ourselves to the cloud, as Ray Kurzweil has suggested? Or, bolstered by yet-to-be-imagined advances in medicine, achieve genuine immortality in Woody Allen's more old-fashioned way, which is to say by not dying? Or by merging our brains with silicon? Nerd Rapture may be near, it may be far, or it may never come. We have no idea.

When Thales investigated electricity in 600 BC, he knew he was onto something, but it would have been impossible to anticipate exactly what; we doubt he dreamt of how electricity would lead to social networking, smart watches, or Wikipedia. It would be arrogant to suppose that we could forecast where AI will be, or the impact it will have in a thousand or even five hundred years.

What we do know is that AI is on its way—and we are best off trying to make sure that whatever comes next is safe, trustworthy, and reliable, directed as much as possible toward helping humanity.

And the best way to make progress toward that goal is to move beyond big data and deep learning alone, and toward a robust new form of AI—carefully engineered, and equipped from the factory with values, common sense, and a deep understanding of the world.

# Acknowledgments

Our goal in this book has been to educate and to challenge, both to explain how AI and machine learning work and how they might be improved. To the extent that we have succeeded, we have been helped immeasurably by colleagues, friends, and family members, often with pressing deadlines of their own. Many, including Mark Achbar, Joey Davis, Annie Duke, Doug Hofstadter, Hector Levesque, Kevin Leyton-Brown, Vik Moharir, Steve Pinker, Philip Rubin, Harry Shearer, Manuela Veloso, Athena Vouloumanos, and Brad Wyble, were kind enough to read and comment on the entire manuscript. Others, including Uri Ascher, Rod Brooks, David Chalmers, Animesh Garg, Yuling Gu, and Cathy O'Neil, gave valuable comments on specific chapters.

We also thank a set of friends and colleagues—Karen Bakker, Leon Bottou, Kyunghyun Cho, Zack Lipton, Missy Cummings, Pedro Domingos, Ken Stanley, Sydney Levine, Omer Levy, Ben Schneiderman, Harry Shearer, and Andrew Sundstrom—for a steady stream of information and pointers. Harry, in particular, never stopped sending us intriguing links, several of which made it into the book.

Maayan Harel's charming and witty drawings greatly enliven the book. We are also grateful to Michael Alcorn, Anish Athalye, Tom Brown, Kevin Eykholt, Kunihiko Fukushima, Gary Lupyan, Tyler

Vigen, and Oriol Vinyals for their gracious permission to use pictures and artwork that they have created; and Steve Pinker and Doug Hofstadter for permission to quote at length from their writings.

We also thank our agent, Daniel Greenberg, who helped connect us with our editor, Edward Kastenmeier, and the team at Pantheon.

Four people especially stand out. Edward Kastenmeier suggested to us a framework that was hugely helpful in organizing the presentation of our arguments, in addition to providing a wealth of brilliant and insightful editing improvements. Steve Pinker, a source of inspiration to Gary for three decades, led us to rethink how we framed the whole book. Annie Duke, fresh from a return to cognitive science after a brief excursion into the world of poker championships, provided fantastic insights into how to better engage lay readers. And Athena Vouloumanos, as she so often has, played two roles, ever-supportive wife to Gary and near-professional editor, reading multiple drafts, every time finding dozens of subtle but powerful ways to radically improve our writing. We are both so very grateful to all.

# Suggested Readings

**AI in General:** The leading textbook for artificial intelligence, and the most comprehensive presentation of the field as a whole, is Stuart Russell and Peter Norvig, *Artificial Intelligence: A Modern Approach.*

A recent online series of articles, "Future of Robotics and Artificial Intelligence," https://rodneybrooks.com/forai-future-of-robotics-and-artificial-intelligence/ by leading roboticist Rodney Brooks (inventor of the Roomba) is very readable and very much in the spirit of our book. Brooks includes a lot of fascinating specifics, both about the practicalities of robotics, and about the history of AI.

**Skepticism About AI:** There have always been those with contrarian views of AI. Early works of this kind include Joseph Weizenbaum's *Computer Power and Human Reason;* and Hubert Dreyfus, *What Computers Can't Do. The AI Delusion,* by Gary Smith; *Artificial Intelligence: Against Humanity's Surrender to Computers,* by Harry Collins; and *Artificial Unintelligence: How Computers Misunderstand the World,* by Meredith Broussard are recent books in a similar vein.

**What's at Stake:** Several important books have been published recently on the risks in AI, short and long term. *Weapons of Math Destruction,* by Cathy O'Neil, and *Automating Inequality: How*

*High-Tech Tools Profile, Police, and Punish the Poor,* by Virginia Eubanks, discuss the potential for social abuse inherent in the use of big data and machine learning by government, insurance companies, employers, and so on.

**Machine Learning and Deep Learning:** The central chapters of *The Master Algorithm: How the Quest for the Ultimate Learning Machine Will Remake Our World,* by Pedro Domingos, are a very readable introduction to machine-learning technologies, with chapters on each of the major approaches to machine learning. *The Deep Learning Revolution,* by Terrence Sejnowski, gives a historical and biographical account. Important recent textbooks in ML include *Machine Learning: A Probabilistic Perspective,* by Kevin Murphy, and *Deep Learning* by Ian Goodfellow, Yoshua Bengio, and Aaron Courville. There are many free machine-learning software libraries and data sets that are available online including Weka Data Mining Software, Pytorch, fast.ai, TensorFlow, Zach Lipton's interactive Jupyter notebooks, and Andrew Ng's popular machine-learning course on Coursera. Guides for these include *Introduction to Machine Learning with Python,* by Andreas Müller and Sarah Guido, and *Deep Learning with Python,* by François Chollet.

**AI Systems That Read:** There is not much written specifically for the lay reader, but the textbooks in this area often include substantial sections that are accessible to the non-expert. The standard textbooks are *Speech and Language Processing,* by Daniel Jurafsky and James H. Martin, and *Foundations of Statistical Natural Language Processing,* by Christopher Manning and Hinrich Schütze. *Introduction to Information Retrieval,* by Christopher Manning, Prabhakar Raghavan, and Hinrich Schütze, is a fine introduction to web search engines and similar programs. As with machine learning, there are software libraries and data sets available online; the most widely used are the Natural Language Toolkit (usually abbreviated NLTK) at https://www.nltk.org and the Stanford Core NLP, at https://stanfordnlp.github.io/CoreNLP/; *Natural Language Processing with Python: Analyzing Text with the Natural Language*

*Toolkit,* by Steven Bird, Ewan Klein, and Edward Loper, is a guide to using NLTK in programs. Douglas Hofstadter's article "The Shallowness of Google Translate" (*The Atlantic,* Jan. 30, 2018) is an enjoyable and insightful analysis of the limitations of current approaches to machine translation.

**Robotics:** Other than the Rodney Brooks online articles mentioned above, there is a dearth of useful popular science writing about robotics. Matthew Mason's fine survey article "Toward Robotic Manipulation" (2018) discusses both biological and robot manipulation. *Modern Robotics: Mechanics, Planning, and Control,* by Kevin Lynch and Frank Park, is an introductory textbook. *Planning Algorithms,* by Steven LaValle, is an overview of high-level planning for robotic motion and manipulation.

**Mind:** Of course, there is no end to the literature here. Particular favorites of ours include *The Language Instinct* and *Words and Rules: The Ingredients of Language,* both by Steven Pinker, for linguistics; *How the Mind Works* and *The Stuff of Thought,* by Pinker, *Kluge,* by Gary Marcus, and *Thinking, Fast and Slow,* by Daniel Kahneman for psychology; and *Brainstorms,* by Daniel Dennett, and *Human Knowledge: Its Scope and Limits,* by Bertrand Russell, for epistemology. Gary's more technical book *The Algebraic Mind,* written in 2001, presages many of the issues that affect contemporary deep learning.

**Commonsense Reasoning:** A recent article by the authors of this book, "Commonsense Reasoning and Commonsense Knowledge in Artificial Intelligence," is similar to chapter 7, but it is longer and includes more detail. *Common Sense, the Turing Test, and the Quest for Real AI,* by Hector Levesque, argues, as we have, that commonsense reasoning is a critical step in achieving genuine intelligence. *Representations of Commonsense Knowledge,* by Ernest Davis, is a textbook on the use of mathematical logic for representing commonsense knowledge. *The Handbook of Knowledge Representation,* edited by Frank van Harmelen, Vladimir Lifschitz, and Bruce Porter, is a useful collection of surveys for more in-depth study. *The*

*Book of Why: The New Science of Cause and Effect,* by Judea Pearl and Dana Mackenzie, discusses automating causal reasoning.

**Trust:** *Moral Machines: Teaching Robots Right from Wrong,* by Wendell Wallach and Colin Allen, and *Robot Ethics: The Ethical and Social Implications of Robotics,* edited by Patrick Lin, Keith Abney, and George Bekey, discuss the problems of instilling a moral sense into robots and AI systems.

*SuperIntelligence: Paths, Dangers, Strategies,* by Nick Bostrom, argues that AI will inevitably undergo a "Singularity" in which it rapidly gains intelligence and passes out of human control. Bostrom describes a variety of scenarios, ranging from dystopian to apocalyptic, for what this would mean for the human race, and discusses the possibility of developing strategies to make sure that AI remains benevolent.

**Future of AI:** Discussions of the long-term impact of AI on human life and society include *Life 3.0: Being Human in the Age of Artificial Intelligence,* by Max Tegmark; *Abundance: The Future Is Better Than You Think,* by Peter Diamandis and Steven Kotler; *Our Final Invention: Artificial Intelligence and the End of the Human Era,* by James Barrat; *Artificial Intelligence: A Futuristic Approach,* by Roman Yampolskiy; and *The Fourth Age: Smart Robots, Conscious Computers, and the Future of Humanity,* by Bryon Reece. *Machines That Think: The Future of Artificial Intelligence,* by Toby Walsh, includes an extensive discussion of the impact of AI in the short- and long-term future, particularly its impact on employment. He also discusses the many different kinds of efforts under way, from research labs to organizations to statements of principles, to ensure that, overall, AI remains safe and beneficent.

# Notes

## 1. MIND THE GAP

3 Marvin Minsky, John McCarthy, and Herb Simon genuinely believed: Minsky, 1967, 2, as quoted in the text. McCarthy: "We think that a significant advance can be made in one or more of these problems if a carefully selected group of scientists work on it together for a summer" in McCarthy, Minsky, Rochester, and Shannon, 1955. Simon, 1965, 96, as quoted in the chapter epigraph.

3 "the problem of artificial intelligence": Minsky, 1967, 2.

3 "surpass native human intelligence": Kurzweil, 2002.

3 "near term AGI": Quoted in Peng, 2018.

4 not everyone is as bullish: Ford, 2018.

4 "autonomous cars [in] the near future": Vanderbilt, 2012.

4 "revolutionize healthcare": IBM Watson Health, 2016.

4 "cognitive systems [could] understand": Fernandez, 2016.

4 "with [recent advances in] cognitive computing": IBM Watson Health, undated.

4 IBM aimed to address problems: IBM Watson Health, 2016.

4 "stop training radiologists": *The Economist,* 2018.

4 M, a chatbot that was supposed: Cade Metz, 2015.

4 Waymo would shortly have driverless cars: Davies, 2017.

4 the bravado was gone: Davies, 2018.

5 widespread recognition that we are at least: Brandom, 2018.

5 MD Anderson Cancer Center shelved: Herper, 2017.

5 "unsafe and incorrect": Ross, 2018.
5 A 2016 project to use Watson: BBC Technology, 2016.
5 "the performance was unacceptable": Müller, 2018.
5 Facebook's M was quietly canceled: Newton, 2018.
5 Eric Schmidt, the former CEO of Google: Zogfarharifard, 2016.
5 "definitely going to rocket us up": Diamandis and Kotler, 2012.
5 "AI is one of the most important": Quoted in Goode, 2018.
5 Google was forced to admit: Simonite, 2019.
5 Bostrom grappled with the prospect: Bostrom, 2014.
5 "human history might go the way": Kissinger, 2018.
6 "summoning the demon": McFarland, 2014.
6 "worse than nukes": D'Orazio, 2014.
6 "the worst event in the history": Molina, 2017.
6 rear-ending parked emergency vehicles: Stewart, 2018; Damiani, 2018.
6 people often overestimate: Zhang and Dafoe, 2019.
6 "Robots Can Now Read Better than Humans": Cuthbertson, 2018.
6 "Computers Are Getting Better than Humans at Reading": Pham, 2018.
7 Stanford Question Answering Dataset: Rajpurkar, Zhang, Lopyrev, and Liang, 2016.
7 "AI that can read a document": Linn, 2018.
8 Facebook introduced a bare-bones proof-of-concept program: Weston, Chopra, and Bordes, 2015.
8 "Facebook Thinks It Has Found": Oremus, 2016.
8 "Facebook AI Software Learns": Rachel Metz, 2015.
9 the public has come to believe: Zhang and Dafoe, 2019.
10 A startup company we are fond of, Zipline: Lardieri, 2018.
10 ImageNet, a library: Deng et al., 2009.
10 chess player AlphaZero: Silver et al., 2018.
10 Google Duplex: Leviathan, 2018.
10 diagnosing skin cancer: Estava et al., 2017.
10 predicting earthquake aftershocks: Vincent, 2018d.
10 detecting credit card fraud: Roy et al., 2018.
10 used in art: Lecoutre, Negrevergne, and Yger, 2017.
10 and music: Briot, Hadjeres, and Pachet, 2017.
10 deciphering speech: Zhang, Chan, and Jaitly, 2017.
10 labeling photos: He and Deng, 2017.
10 organizing people's news feeds: Hazelwood et al., 2017.

10 to identify plants: Matchar, 2017.
10 enhance the sky in your photos: Hall, 2018.
10 colorize old black-and-white pictures: He et al., 2018.
10 epic battle for talent: Metz, 2017.
11 sold out in twelve minutes: Falcon, 2018.
11 France, Russia, Canada, and China: Fabian, 2018.
11 China alone is planning: Herman, 2018.
11 McKinsey Global Institute estimates: Bughin et al., 2018.
11 performs two types of analysis: Kintsch and van Dijk, 1978; Rayner, Pollatsek, Ashby, and Clifton, 2012.
12 they can't understand the news: Marcus and Davis, 2018.
12 The technology just isn't mature yet: Romm, 2018.
13 "Taking me from Cambridge": Lippert, Gruley, Inoue, and Coppola, 2018; Romm, 2018; Marshall, 2017.
13 Google Duplex: Statt, 2018.
14 could handle just three things: Leviathan, 2018.
14 nothing but restaurant reservations: Callahan, 2019.
15 Thirty thousand people a year: Wikipedia, "List of Countries by Traffic-Related Death Rate."
17 narrow AI tends to get flummoxed: Marcus, 2018a; Van Horn and Perona, 2017.
19 central principle of social psychology: Ross, 1977.
19 a chatbot called Eliza: Weizenbaum, 1966.
19 "People who knew very well": Weizenbaum, 1965, 189–90.
19 In 2016, a Tesla owner: Levin and Woolf, 2016.
19 The car appears to have warned him: Fung, 2017.
21 success in closed-world tasks: McClain, 2011.
21 "typically perform only under": Missy Cummings, email to authors, September 22, 2018.
21 *Dota 2:* Vincent, 2018.
21 *Starcraft 2:* AlphaStar Team, 2019.
21 almost none of its training: Vinyals, 2019.
23 correctly labeled by a highly touted: Vinyals, Toshev, Bengio, and Erhan, 2015.
23 "refrigerator filled with lots of food and drinks": Vinyals, Toshev, Bengio, and Erhan, 2015.
24 Teslas repeatedly crashing into parked fire engines: Stewart, 2018.
24 mislead people with statistics: Huff, 1954.

24 suitable for some problems but not others: Müller, 2018.
25 what he thought AI couldn't do: Dreyfus, 1979.

## 2. WHAT'S AT STAKE

27 "A lot can go wrong": O'Neil, 2017.
27 Xiaoice: Thompson, 2016; Zhou, Gao, Li, and Shum, 2018.
27 the project was canceled: Bright, 2016. The Tay debacle has been set to verse in Davis, 2016b.
27 Alexas that spooked their owners: Chokshi, 2018.
27 iPhone face-recognition systems: Greenberg, 2017.
27 Poopocalypse: Solon, 2016.
27 hate speech detectors: Matsakis, 2018.
27 job candidate systems: Dastin, 2018.
27 ludicrous conspiracy theories: Porter, 2018; Harwell and Timberg, 2019.
27 sent a jaywalking ticket: Liao, 2018.
28 backing out of its owners' garage: Harwell, 2018.
28 robotic lawnmowers have maimed: Parker, 2018.
28 iPhone that autocorrects: http://autocorrectfailness.com/autocorrect-fail-ness-16-im-gonnacrapholescreenshots/happy-birthday-dead-papa/.
28 A report from the group AI Now: Campolo, 2017.
28 Flash crashes on Wall Street: Seven, 2014.
28 an Alexa recorded a conversation: Canales, 2018.
28 multiple automobile crashes: Evarts, 2016; Fung, 2017.
30 "[T]he scenario": Pinker, 2018.
31 the Amelia Bedelia problem: Parish, 1963.
32 seriously, but not always literally: Zito, 2016.
32 a pedestrian walkway in Miami: Mazzei, Madigan, and Hartocollis, 2016.
32 Sherry Turkle has pointed out: Turkle, 2017.
32 To take a more subtle example: Coldewey, 2018.
33 Machine-translation systems trained on legal documents: Koehn and Knowles, 2017.
33 Voice-recognition systems: Huang, Baker, and Reddy, 2014.
33 flamed out when the colors were reversed: Hosseini, Xiao, Jaiswal, and Poovendran, 2017.
33 there are blue stop signs in Hawaii: Lewis, 2016.

33 Judy Hoffman has shown: Hoffman, Wang, Yu, and Darrell, 2016.
33 Latanya Sweeney discovered: Sweeney, 2013.
33 in 2015, Google Photos mislabeled: Vincent, 2018a.
33 "Professional hair style for work": O'Neil, 2016b.
34 In 2018, Joy Buolamwini: Buolamwini and Gebru, 2018.
34 IBM was the first to patch: Vincent, 2018b.
34 Microsoft swiftly followed suit: Corbett and Vaniar, 2018.
34 closer to half of professors are women: NCES, 2019.
34 recruiting system . . . was so problematic: Dastin, 2018.
34 The data sets used for training: Lashbrook, 2018. In fairness, the problems with bias in the medical literature toward studies involving white, male subjects long predate AI medicine.
34 measurably less reliable: Wilson, Hoffman, and Morgenstern, 2019.
35 significant fraction of the texts on the web: Venugopal, Uszkoreit, Talbot, Och, and Ganitkevitch, 2011.
35 a lot of allegedly high-quality human-labeled data: Dreyfuss, 2018.
35 people succeeded in getting Google Images: Hayes, 2018.
35 Sixteen years earlier: Wilson, 2011.
36 even if a program is written: O'Neil, 2016a.
37 the company removed this as a criterion: O'Neil, 2016a, 119.
37 DeepMind researcher Victoria Krakovna: Krakovna, 2018.
37 A soccer-playing robot: Ng, Harada, and Russell, 1999.
37 A robot that was supposed to learn: Amodei, Christiano, and Ray, 2017.
38 An unambitious AI tasked: Murphy, 2013.
38 a dairy company hired: Witten and Frank, 2000, 179–80.
38 Stalkers have begun using: Burns, 2017.
38 spammers have used: Hines, 2007.
38 There is little doubt: Efforts by the AI research community to promote a ban on AI-powered autonomous weapons are reviewed in Sample, 2017; and Walsh, 2018. See, for instance, Future of Life Institute, 2015.
38 "When a very efficient technology": Eubanks, 2018, 173.
39 IBM, for example, managed to fix: Vincent, 2018b.
39 Google solved its gorilla challenge: Vincent, 2018a.

## 3. DEEP LEARNING, AND BEYOND

41 "knowledge-based" approach: Davis and Lenat, 1982; Newell, 1982.
42 *machine learning*: Mitchell, 1997.

42 Frank Rosenblatt built a "neural network": Rosenblatt, 1958.
42 reported in *The New York Times*: *New York Times,* 1958.
42 Francis Crick noted; Crick, 1989.
43 the dark days of the 1990s and 2000s: e.g., Hinton, Sejnowski, and Poggio, 1999; Arbib, 2003.
43 special piece of hardware known as a GPU: Barlas, 2015.
43 applied to neural networks since the early 2000s: Oh and Jung, 2004.
43 A revolution came in 2012: Krizhevsky, Sutskever, and Hinton, 2012.
43 Hinton's team scored 84 percent correct: Krizhevsky, Sutskever, and Hinton, 2012.
43 reached 98 percent: Gershgorn, 2017.
44 Hinton and some grad students formed a company: McMillan, 2013.
44 bought a startup called DeepMind: Gibbs, 2014.
46 subjects of news articles: Joachims, 2002.
46 structures of proteins: Hua and Sun, 2001.
46 Probabilistic models: Murphy, 2012.
46 vital for the success of IBM's Watson: Ferrucci et al., 2010.
46 genetic algorithms have been used: Linden, 2002.
46 playing video games: Wilson, Cussat-Blanc, Lupa, and Miller, 2018, 5.
46 Pedro Domingos's book: Domingos, 2015.
47 traffic routing algorithms used by Waze: Bardin, 2018.
47 used a mix of classical AI techniques: Ferrucci et al., 2010.
48 more than $80 billion: Garrahan, 2017.
48 a set of experiments in the 1950s: Hubel and Wiesel, 1962.
48 Neocognitron: Fukushima and Miyake, 1982.
48 Later books by Jeff Hawkins and Ray Kurzweil: Hawkins and Blakeslee, 2004; Kurzweil, 2013.
49 backbone of deep learning: LeCun, Hinton, and Bengio, 2015.
50 In their influential 1969 book: Minsky and Papert, 1969.
51 Over the subsequent two decades: A number of people have been credited with the independent discovery of versions of backpropagation, including Henry Kelly in 1960, Arthur Bryson in 1961, Stuart Dreyfus in 1962, Bryson and Yu-Chi Ho in 1969, Seppo Linnainman in 1970, Paul Werbos in 1974, Yann LeCun in 1984, D. B. Parker in 1985, and David Rumelhart, Geoffrey Hinton, and Ronald Williams in 1986. See Wikipedia, "Backpropagation"; Russell and Norvig, 2010, 761; and LeCun, 2018.

51 *backpropagation*: Rumelhart, Hinton, and Williams, 1986.
52 A technique called *convolution*: LeCun and Bengio, 1995.
52 It would have taken: Nandi, 2015.
53 some important technical tweaks: Srivastava, Hinton, Krizhevsky, Sutskever, and Salakhutdinov, 2014; Glorot, Bordes, and Bengio, 2011.
53 sometimes more than a hundred: He, Zhang, Ren, and Sun, 2016.
53 deep learning has radically improved: Lewis-Krauss, 2016.
54 yielded markedly better translations: Bahdanau, Cho, and Bengio, 2014.
54 transcribe speech and label photographs: Zhang, Chan, and Jaitly, 2017; He and Deng, 2017.
54 turn your landscape into a Van Gogh: Gatys, Ecker, and Bethge, 2016.
54 colorizing old pictures: Iizuka, Simo-Serra, and Ishikawa, 2016.
54 "unsupervised learning": Chintala and LeCun, 2016.
55 Atari video games: Mnih et al., 2015.
55 later on Go: Silver et al., 2016; Silver et al., 2017.
55 "If a typical person": Ng, 2016.
55 his research a dozen years earlier: Marcus, 2001.
55 "Realistically, deep learning": Marcus, 2012b.
56 three core problems: Marcus, 2018a.
56 AlphaGo required 30 million games: Silver, 2016. Slide 18.
57 why neural networks work: Bottou, 2018.
58 *A woman talking on a cell phone*: Rohrbach, Hendricks, Burns, Darrell, and Saenko, 2018.
59 a three-dimensional turtle: Athalye, Engstrom, Ilyas, and Kwok, 2018.
59 tricked neural nets into thinking: Karmon, Zoran, and Goldberg, 2018.
59 psychedelic pictures of toasters: Brown, Mané, Roy, Abadi, and Gilmer, 2017.
60 deliberately altered stop sign: Evtimov et al., 2017.
60 twelve different tasks: Geirhos et al., 2018.
60 deep learning has trouble recognizing: Alcorn et al., 2018.
61 systems that work on the SQuAD task: Jia and Liang, 2017.
61 Another study showed how easy: Agrawal, Batra, and Parikh, 2016.
62 translate that from Yoruba: Greg, 2018. This was widely reported, and verified by the authors.

62 deep learning just ain't that deep: Marcus, 2018a. Alex Irpan, a software engineer at Google, has made similar points with respect to deep reinforcement learning: Irpan, 2018.

62 DeepMind's entire system falls apart: Kansky et al., 2017.

62 tiny bits of noise shattered performance: Huang, Papernot, Goodfellow, Duan, and Abbeel, 2017.

62 "deep [neural networks] tend to learn": Jo and Bengio, 2017.

62 nowhere close to being a reality: Wiggers, 2018.

63 "long tail" problem: Piantadosi, 2014; Russell, Torralba, Murphy, and Freeman, 2008.

63 this picture: https://pxhere.com/en/photo/1341079.

64 2013 annual list of breakthrough technologies: Hof, 2013.

66 In the immortal words of Law 31: Akin's Laws of Spacecraft Design. https://spacecraft.ssl.umd.edu/akins_laws.html.

## 4. IF COMPUTERS ARE SO SMART, HOW COME THEY CAN'T READ?

68 Google Talk to Books: Kurzweil and Bernstein, 2018.

68 "Google's astounding new search tool": Quito, 2018.

69 encoding the meanings of sentences: An earlier technique, Latent Semantic Analysis, also converted natural language expressions into vectors. Deerwester, Dumais, Furnas, Landauer, and Harshman, 1990.

69 Yet when we asked: Experiment carried out by the authors, April 19, 2018.

71 "Almanzo turned to Mr. Thompson": Wilder, 1933.

74 deep, broad, and flexible: Dyer, 1983; Mueller, 2006.

74 "Today would have been Ella Fitzgerald's": Levine, 2017.

76 getting machines to understand stories: Norvig, 1986.

76 how machines could use "scripts": Schank and Abelson, 1977.

77 PageRank algorithm: Page, Brin, Motwani, and Winograd, 1999.

78 "What is the capital of Mississippi?": This and "What is 1.36 euros in rupees?" were experiments carried out by the authors, May 2018.

78 "Who is currently on the Supreme Court?": Experiments carried out by the authors, May 2018.

79 "When was the first bridge ever built?": Experiment carried out by the authors, August 2018. The passage that Google retrieved is from Ryan, 2001–2009.

79 off by thousands of years: The Arkadiko bridge, in Greece, built around 1300 BC, is still standing. But that is a sophisticated stone arch bridge; human beings undoubtedly constructed more primitive, and less permanent, bridges for centuries or millennia before that.

80 directions to the nearest airport: Bushnell, 2018.

80 On a recent drive that one of us took: Experiment carried out May 2018.

81 "the world's first computational knowledge engine": WolframAlpha Press Center, 2009. The home page for WolframAlpha is https://www.wolframalpha.com.

81 But the limits of its understanding: Experiments with WolframAlpha carried out by the authors, May 2018.

82 titles of Wikipedia pages: Chu-Carroll et al., 2012.

82 When we looked recently: Search carried out May 2018. The demo of IBM Watson Assistant is at https://watson-assistant-demo.ng.bluemix.net/.

84 As the *New York Times Magazine* article: Lewis-Krauss, 2016.

85 If you give Google Translate the French sentence: Experiment carried out by the authors, August 2018. Ernest Davis maintains a website with a small collection of mistakes made by leading machine translation programs on sentences that should be easy: https://cs.nyu.edu/faculty/davise/papers/GTFails.html.

85 "We humans know all sorts of things": Hofstadter, 2018.

87 when we asked Google Translate: Experiment conducted by the authors, August 2018. Google Translate also makes the corresponding mistake in translating the same sentence into German, Spanish, and Italian.

89 build up a *cognitive model*: Kintsch and Van Dijk, 1978.

89 object file: Kahneman, Treisman, and Gibbs, 1992.

## 5. WHERE'S ROSIE?

95 sometimes even falling over: IEEE Spectrum, 2015, 0:30.

95 a robot opening one particular doorknob: Glaser, 2018.

96 leave a trail of banana peels: Boston Dynamics, 2016, 1:25–1:30.

98 pet-like home robots: For instance, Sony released an updated version of its robot dog Aibo in spring 2018: Hornyak, 2018.

98 "driverless" suitcases: Brady, 2018.

99 tiny amounts of computer hardware: The first generation Roomba, re-

leased in 2002, used a computer with 256 bytes of writable memory. That's not a typo. That's about one billionth as much memory as an iPhone. Ulanoff, 2002.

99 robots that can safely wander the halls: Veloso, Biswas, Coltin, and Rosenthal, 2015.

100 a robotic butler: Lancaster, 2016.

100 SpotMini, a sort of headless robotic dog: Gibbs, 2018.

100 Atlas robot: Boston Dynamics, 2017; Boston Dynamics, 2018a; Harridy, 2018.

100 that parkour video: CNBC, 2018.

100 WildCat: Boston Dynamics, 2018c.

100 BigDog: Boston Dynamics, 2018b.

100 MIT's Sangbae Kim: Kim, Laschi, and Trimmier, 2013.

101 Robots from the iRobot company: Brooks, 2017b.

103 OODA loop: Wikipedia, "OODA Loop."

103 accurate to within about ten feet: Kastranakes, 2017.

103 Simultaneous Localization And Mapping: Thrun, 2007.

105 devise a complex plan: Mason, 2018.

105 making good progress on motor control: OpenAI blog, 2018; Berkeley CIR, 2018.

109 "too much automation": Allen, 2018.

110 PR2 fetching a beer from a refrigerator: Willow Garage, 2010.

110 Even the fridge was specially arranged: Animesh Garg, email to authors, October 24, 2018.

114 its first fatal accident: Evarts, 2016.

## 6. INSIGHTS FROM THE HUMAN MIND

116 "causal entropic forces": Wissner-Gross and Freer, 2013.

116 "walk upright, use tools": Bot Scene, 2013.

116 "broad applications": Wissner-Gross, 2013.

116 "figured out a 'law'": Ball, 2013a. Ball somewhat revised his views in a later blog, Ball, 2013b.

116 TED gave Wissner-Gross a platform: Wissner-Gross, 2013.

117 "In suggesting that causal entropy": Marcus and Davis, 2013.

117 gone on to other projects: See Wissner-Gross's website: http://www.alexwg.org.

117 behaviorism became all the rage: Watson, 1930; Skinner, 1938.

117 induce precise, mathematical causal laws: Skinner, 1938.
118 "there is no one way the mind works": Firestone and Scholl, 2016.
118 Humans are flawed in many ways: Marcus, 2008.
119 roughly 86 billion neurons: Herculano-Houzel, 2016; Marcus and Freeman, 2015.
119 trillions of synapses: Kandel, Schwartz, and Jessell, 1991.
119 hundreds of distinct proteins: O'Rourke, Weiler, Micheva, and Smith, 2012.
119 150 distinctly identifiable brain areas: Amunts and Zilles, 2015.
119 a vast and intricate web of connections: Felleman and van Essen, 1991; Glassert et al., 2016.
119 "Unfortunately, nature seems unaware": Ramón y Cajal, 1906.
120 book review written in 1959: Chomsky, 1959.
120 an effort to explain human language: Skinner, 1957.
122 hand-crafted machinery: Silver et al., 2016; Marcus, 2018b.
122 "The fundamental challenges": Geman, Bienenstock, and Doursat, 1992.
123 Kahneman divides human cognitive process: Kahneman, 2011.
124 the terms *reflexive* and *deliberative*: Marcus, 2008.
124 "society of mind": Minsky, 1986, 20.
124 Howard Gardner's ideas: Gardner, 1983.
124 Robert Sternberg's triarchic theory: Sternberg, 1985.
124 evolutionary and developmental psychology: Barlow, Cosmides, and Tooby, 1996; Marcus, 2008; Kinzler and Spelke, 2007.
124 requires a different subset of our brain resources: Braun et al., 2015; Preti, Bolton, and de Ville, 2017.
125 Nvidia's 2016 model of driving: Bojarski et al., 2016.
125 training end-to-end from pixels: Mnih et al., 2015.
125 could not get a similar approach to work for Go: Silver et al., 2016.
126 fruit flies of linguistics: Pinker, 1999.
127 Part of Gary's PhD work: Marcus et al., 1992.
127 "true intelligence is a lot more": Heath, 2018.
128 The essence of language, for Chomsky: Chomsky, 1959.
128 "thought vectors": Devlin, 2015.
128 Word2Vec: Mikolov, Sutskever, Chen, Corrado, and Dean, 2013.
129 product search on Amazon: Ping et al., 2018.
129 "If you take the vector": Devlin, 2015.

132 If they can't capture individual words: These and other limitations of word embeddings are discussed in Levy, in preparation.

132 "You can't cram the meaning": Mooney is quoted with expletives deleted in Conneau et al., 2018.

132 Take a look at this picture: Lupyan and Clark, 2015.

133 One classic experiment: Carmichael, Hogan, and Walter, 1932.

134 a vision system was fooled: Vondrick, Khosla, Malisiewicz, and Torralba, 2012.

135 If we run these by themselves: Experiment carried out by the authors on Amazon Web Services, August 2018.

136 language tends to be underspecified: Piantadosi, Tily, and Gibson, 2012.

137 you can easily imagine: Rips, 1989.

138 a robin is a prototypical bird: Rosch, 1973.

138 the Yale psychologist Frank Keil: Keil, 1992.

138 concepts that are embedded in theories: Murphy and Medin, 1985; Carey, 1985.

139 a rich understanding of causality: Pearl and MacKenzie, 2018.

141 Vigen compiled a whole book: Vigen, 2015.

142 But to the deep learning system: Pinker, 1997; Marcus, 2001.

143 Mendel himself was initially ignored: Judson, 1980.

143 Individual genes are in fact levers: Marcus, 2004.

144 Piaget's questions: Piaget, 1928.

144 but the answers he proposed: Gelman and Baillargeon, 1983; Baillargeon, Spelke, and Wasserman, 1985.

145 humans are likely born understanding: Spelke, 1994; Marcus, 2018b.

145 as Kant argued two centuries earlier: Kant, 1751/1998.

145 some aspects of language are also: Pinker, 1994.

145 expectations about what language might sound like: Shultz and Vouloumanos, 2010.

145 the results haven't been nearly as impressive: Hermann et al., 2017.

146 "without human knowledge": Silver et al., 2017.

146 nothing intrinsic to do with deep learning: Marcus, 2018b.

146 The claim that human knowledge: Marcus, 2018b.

147 "We need a new generation of AI researchers": Darwiche, 2018.

147 LeCun argued forcefully: LeCun et al., 1989.

## 7. COMMON SENSE, AND THE PATH TO DEEP UNDERSTANDING

150 first started calling attention to it: McCarthy, 1959.

150 ten facts that NELL had recently learned: These results are from a test of NELL carried out by the authors on May 28, 2018.

151 ConceptNet: Havasi, Pustejovsky, Speer, and Lieberman, 2009.

151 The English sentences are then automatically converted: Singh et al., 2002.

152 "Artificial Intelligence Meets Natural Stupidity": McDermott, 1976.

153 VirtualHome: Puig et al., 2018. The VirtualHome project can be found at https://www.csail.mit.edu/research/virtualhome-representing-activities-programs.

153 Schank's work on scripts: Schank and Abelson, 1977.

154 but not in others: Similar objections were raised in Dreyfus, 1979.

154 The work has been both painstaking and difficult: Davis, 2017 is a recent survey of this work. Davis, 1990 and van Harmelen, Lifschitz, and Porter, 2008 are earlier book-length studies.

154 The largest effort in the field: The CYC project was announced in Lenat, Prakash, and Shepherd, 1985. A book-length progress report was published in 1990: Lenat and Guha, 1990. No comprehensive account has been published since.

154 millions of carefully encoded facts: Matuszek et al., 2005.

155 External articles written about it: Conesa, Storey, and Sugumaran, 2010.

155 taxonomy, the kind of categorization: Collins and Quillian, 1969.

156 WordNet: Miller, 1995.

156 the medical taxonomy SNOMED: Schulz, Suntisrivaraporn, Baader, and Boeker, 2009.

157 many other taxonomies are not: One attempt to deal with vague entities and relations is fuzzy logic, developed by Lotfi Zadeh: Zadeh, 1987.

157 it's hard to define: Wittgenstein, 1953.

158 a lot of what you need to know: Woods, 1975; McDermott, 1976.

161 some alternative that does similar work: For example, it is possible to define variants of semantic networks whose meaning is as precisely defined as logical notation. Brachman and Schmolze, 1989; Borgida and Sowa, 1991.

161 frameworks for many different aspects: Davis, 2017.
162 *time, space,* and *causality* are fundamental: Kant, 1751/1998. Steven Pinker argues for a similar view in *The Stuff of Thought:* Pinker, 2007.
168 The ruthless attorney Rosalind Shays: Pinker, 1997, 314.
171 the evolution of galaxies: Benger, 2008.
171 the flow of blood cells: Rahimian et al., 2010.
171 the aerodynamics of helicopters: Padfield, 2008.
171 simulations just won't work: The limits of simulation for AI is discussed at length in Davis and Marcus, 2016.
172 SpotMini: Boston Dynamics, 2016.
173 "reality gap": Mouret and Chatzilygeroudis, 2017.
175 not every inference: Sperber and Wilson, 1986.
175 "the frame problem": Pylyshyn, 1987.
175 automated reasoning: Lifschitz, Morgenstern, and Plaisted, 2008.
176 "All human knowledge is uncertain": Russell, 1948, 307.
178 "core" systems that Spelke has emphasized: Davis, 1990; van Harmelen, Lifschitz, and Porter, 2008.

## 8. TRUST

181 as long as any one of the five was still running: Tomayko, 1988, 100.
181 Elon Musk claimed for years: Hawkins, 2018.
182 San Francisco's cable cars: Cable Car museum, undated.
182 "white hat hackers" were able: Greenberg, 2015.
183 Yet it is fairly easy to block or spoof: Tullis, 2018.
183 the Russian government has hacked: Sciutto, 2018.
183 "a perfect target for cybercriminals": Mahairas and Beshar, 2018.
184 "The problem of [combining] traditional software": Léon Bottou email to the authors, July 19, 2018.
184 the Turing test: Turing, 1950.
184 not particularly useful: Hayes and Ford, 1995.
185 alternatives to the Turing test: Marcus, Rossi, and Veloso, 2016. See also Reddy, Chen, and Manning, 2018; Wang et al., 2018; and the Allen Institute for AI website, https://allenai.org/.
185 language comprehension: Levesque, Davis, and Morgenstern, 2012.
185 inferring physical and mental states: Rashkin, Chap, Allaway, Smith, and Choi, 2018.
185 understanding YouTube videos: Paritosh and Marcus, 2016.

185 elementary science: Schoenik, Clark, Tafjord, Turney, and Etzioni, 2016; Davis, 2016a.

185 robotic abilities: Ortiz, 2016.

185 transfer those skills to other games: Chaplot, Lample, Sathyendra, and Salakhudinov, 2016.

187 Static Driver Verifier: Wikipedia, "Driver Verifier."

187 computerized control program for the Airbus: Souyris, Wiels, Delmas, and Delseny, 2009.

187 verify that the collision avoidance programs: Jeannin et al., 2015.

187 two fatal accidents involving the Boeing 737 Max: Levin and Suhartono, 2019.

188 "move fast and break things": Baer, 2014.

188 "technical debt": Sculley et al., 2014.

191 hundreds of millions of numerical parameters: See, for example, Vaswani et al., 2017, table 3, or Canziani, Culurciello, and Paszke, 2017, figure 2.

191 "explainable AI": Gunning, 2017; Lipton, 2016.

193 "Three Laws of Robotics": Asimov, 1942.

194 What kind of harm or injury: Leben, 2018.

194 moral dilemmas: Wallach and Allen, 2010.

194 the one that Gary introduced: Marcus, 2012a.

195 "No general code of ethics": Sartre, 1957.

196 Bostrom's widely discussed paper-clip example: This was introduced in Bostrom, 2003. It has since then been extensively discussed by many writers, in particular by Nick Bostrom, Eliezer Yudkowsky, and their collaborators. Our discussion here is based primarily on Bostrom, 2014; Yudkowsky, 2011; Bostrom and Yudkowsky, 2014; and Soares, Fallenstein, Armstrong, and Yudkowsky, 2015.

196 "The AI does not hate you": Yudkowsky, 2011.

196 "summoning the demon": McFarland, 2014.

197 unaware of the consequences of its actions: Similar arguments are presented in Pinker, 2018, and in Brooks, 2017c.

197 people who expect AIs to be harmless: Yudkowsky, 2011.

## EPILOGUE

200 using deep learning to track wildlife: Norouzzadeh et al., 2018.

200 predict aftershocks of earthquakes: Vincent, 2018d.

201 at a crucial moment Rick Deckard: Harford, 2018.

202 "PowerPoint annoyances": Swinford, 2006.
205 if Peter Diamandis is right: Diamandis and Kotler, 2012.
205 the vision of Oscar Wilde: Wilde, 1891.

## SUGGESTED READINGS

210 Weka Data Mining Software: https://www.cs.waikato.ac.nz/ml/weka/.
210 Pytorch: https://pytorch.org.
210 fast.ai: https://www.fast.ai/.
210 TensorFlow: https://www.tensorflow.org/.
210 Zach Lipton's interactive Jupyter notebooks: https://github.com/zackchase/mxnet-the-straight-dope.
210 Andrew Ng's popular machine-learning course: https://www.coursera.org/learn/machine-learning.

# Bibliography

*Links for many of these references may be found at Rebooting.AI .com.*

Agrawal, Aishwarya, Dhruv Batra, and Devi Parikh. 2016. "Analyzing the behavior of visual question answering models." *arXiv preprint arXiv:1606.07356*. https://arxiv.org/abs/1606.07356.

Alcorn, Michael A., Qi Li, Zhitao Gong, Chengfei Wang, Long Mai, Wei-Shinn Ku, and Anh Nguyen. 2018. "Strike (with) a pose: Neural networks are easily fooled by strange poses of familiar objects." *arXiv preprint arXiv:1811.11553*. https://arxiv.org/abs/1811.11553.

Allen, Tom. 2018. "Elon Musk admits 'too much automation' is slowing Tesla Model 3 production." *The Inquirer*. April 16, 2018. https://www.theinquirer.net/inquirer/news/3030277/elon-musk-admits-too-much-automation-is-slowing-tesla-model-3-production.

AlphaStar Team. 2019. "AlphaStar: Mastering the real-time strategy game StarCraft II." https://deepmind.com/blog/alphastar-mastering-real-time-strategy-game-starcraft-ii/.

Amodei, Dario, Paul Christiano, and Alex Ray. 2017. "Learning from human preferences." *OpenAI Blog*. June 13, 2017. https://blog.openai.com/deep-reinforcement-learning-from-human-preferences/.

Amunts, Katrin, and Karl Zilles. 2015. "Architectonic mapping of the human brain beyond Brodmann." *Neuron* 88(6): 1086–1107. https://doi.org/10.1016/j.neuron.2015.12.001.

Arbib, Michael. 2003. *The Handbook of Brain Theory and Neural Networks*. Cambridge, MA: MIT Press.

Asimov, Isaac. 1942. "Runaround." *Astounding Science Fiction*. March 1942. Included in Isaac Asimov, *I, Robot*, Gnome Press, 1950.

Athalye, Anish, Logan Engstrom, Andrew Ilyas, and Kevin Kwok. 2018. "Synthesizing robust adversarial examples." *Proc. 35th Intl. Conf. on Machine Learning.* http://proceedings.mlr.press/v80/athalye18b/athalye18b.pdf.

Baer, Drake. 2014. "Mark Zuckerberg explains why Facebook doesn't 'move fast and break things' anymore." *Business Insider,* May 2, 2014. https://www.businessinsider.com/mark-zuckerberg-on-facebooks-new-motto-2014-5.

Bahdanau, Dzmitry, Kyunghyun Cho, and Yoshua Bengio. 2014. "Neural machine translation by jointly learning to align and translate." *arXiv preprint arXiv:1409.0473*. https://arxiv.org/abs/1409.0473.

Baillargeon, Renee, Elizabeth S. Spelke, and Stanley Wasserman. 1985. "Object permanence in five-month-old infants." *Cognition* 20(3): 191–208. https://doi.org/10.1016/0010-0277(85)90008-3.

Ball, Philip. 2013a. "Entropy strikes at the *New Yorker*." *Homunculus* blog. May 9, 2013. http://philipball.blogspot.com/2013/05/entropy-strikes-at-new-yorker.html.

Ball, Philip. 2013b. "Stuck in the middle again." *Homunculus* blog. May 16, 2013. http://philipball.blogspot.com/2013/05/stuck-in-middle-again.html.

Bardin, Noam. 2018. "Keeping cities moving—how Waze works." *Medium.com*. April 12, 2018. https://medium.com/@noambardin/keeping-cities-moving-how-waze-works-4aad066c7bfa.

Barlas, Gerassimos. 2015. *Multicore and GPU Programming.* Amsterdam: Morgan Kaufmann.

Barlow, Jerome, Leda Cosmides, and John Tooby. 1996. *The Adapted Mind: Evolutionary Psychology and the Generation of Culture*. Oxford: Oxford University Press.

Barrat, James. 2013. *Our Final Invention: Artificial Intelligence and the End of the Human Era*. New York: Thomas Dunne Books/St. Martin's Press.

BBC Technology. 2016. "IBM AI system Watson to diagnose rare diseases in Germany." October 18, 2016. https://www.bbc.com/news/technology-37653588.

Benger, Werner. 2008. "Colliding galaxies, rotating neutron stars and merging black holes—visualizing high dimensional datasets on arbitrary meshes." *New Journal of Physics* 10(12): 125004. http://dx.doi.org/10.1088/1367-2630/10/12/125004.

Berkeley CIR. 2018. Control, Intelligent Systems and Robotics (CIR). Website. https://www2.eecs.berkeley.edu/Research/Areas/CIR/.

Bird, Steven, Ewan Klein, and Edward Loper. 2009. *Natural Language Processing with Python: Analyzing Text with the Natural Language Toolkit.* Cambridge, MA: O'Reilly Pubs.

Bojarski, Mariusz, et al. 2016. "End-to-end deep learning for self-driving cars." *NVIDIA Developer Blog.* https://devblogs.nvidia.com/deep-learning-self-driving-cars/.

Borgida, Alexander, and John Sowa. 1991. *Principles of Semantic Networks: Explorations in the Representation of Knowledge.* San Mateo, CA: Morgan Kaufmann.

Boston Dynamics. 2016. "Introducing SpotMini." Video. https://www.youtube.com/watch?v=tf7IEVTDjng.

Boston Dynamics. 2017. *What's New, Atlas?* Video. https://www.youtube.com/watch?v=fRj34o4hN4I.

Boston Dynamics. 2018a. "Atlas: The world's most dynamic humanoid." https://www.bostondynamics.com/atlas.

Boston Dynamics. 2018b. "BigDog: The first advanced rough-terrain robot." https://www.bostondynamics.com/bigdog.

Boston Dynamics. 2018c. "WildCat: The world's fastest quadruped robot." https://www.bostondynamics.com/wildcat.

Bostrom, Nick. 2003. "Ethical issues in advanced artificial intelligence." *Science Fiction and Philosophy: From Time Travel to Superintelligence*, edited by Susan Schneider. 277–284. Hoboken, NJ: Wiley and Sons.

Bostrom, Nick. 2014. *SuperIntelligence: Paths, Dangers, Strategies.* Oxford: Oxford University Press.

Bostrom, Nick, and Eliezer Yudkowsky. 2014. "The ethics of artificial intelligence." In *The Cambridge Handbook of Artificial Intelligence,* edited by Keith Frankish and William Ramsey, 316–334. Cambridge: Cambridge University Press.

Bot Scene 2013. "Entropica claims 'powerful new kind of AI.'" *Bot Scene* blog. May 11, 2013. https://botscene.net/2013/05/11/entropica-claims-powerful-new-kind-of-ai/comment-page-1/.

Bottou, Léon. 2018. Foreword. In Marvin Minsky and Seymour Papert, *Perceptrons: An Introduction to Computational Geometry.* Reissue of the 1988 expanded edition, with a new foreword by Léon Bottou. Cambridge, MA: MIT Press.

Brachman, Ronald J., and James G. Schmolze. "An overview of the KL-ONE knowledge representation system." In *Readings in Artificial Intelligence and Databases,* edited by John Myopoulos and Michael Brodie, 207–230. San Mateo, CA: Morgan Kaufmann, 1989.

Brady, Paul. 2018. "Robotic suitcases: The trend the world doesn't need." *Condé Nast Traveler*. January 10, 2018. https://www.cntraveler.com/story/robotic-suitcases-the-trend-the-world-doesnt-need.

Brandom, Russell. 2018. "Self-driving cars are headed toward an AI roadblock." *The Verge*. July 3, 2018. https://www.theverge.com/2018/7/3/17530232/self-driving-ai-winter-full-autonomy-waymo-tesla-uber.

Braun, Urs, Axel Schäfer, Henrik Walter, Susanne Erk, Nina Romanczuk-Seiferth, Leila Haddad, Janina I. Schweiger, et al. 2015. "Dynamic reconfiguration of frontal brain networks during executive cognition in humans." *Proceedings of the National Academy of Sciences* 112(37): 11678–11683. https://doi.org/10.1073/pnas.1422487112.

Bright, Peter. 2016. "Tay, the neo-Nazi millennial chatbot, gets autopsied." *Ars Technica*. May 25, 2016. https://arstechnica.com/information-technology/2016/03/tay-the-neo-nazi-millennial-chatbot-gets-autopsied/.

Briot, Jean-Pierre, Gaëtan Hadjeres, and François Pachet. 2017. Deep learning techniques for music generation—a survey. *arXiv preprint arXiv:1709.01620*. https://arxiv.org/abs/1709.01620.

Brooks, Rodney. 2017a. "Future of robotics and artificial intelligence." https://rodneybrooks.com/forai-future-of-robotics-and-artificial-intelligence/.

Brooks, Rodney. 2017b. "Domo Arigato Mr. Roboto." http://rodneybrooks.com/forai-domo-arigato-mr-roboto/.

Brooks, Rodney. 2017c. "The seven deadly sins of predicting AI." http://rodneybrooks.com/the-seven-deadly-sins-of-predicting-the-future-of-ai/.

Broussard, Meredith. 2018. *Artificial Unintelligence: How Computers Misunderstand the World*. Cambridge, MA: MIT Press.

Brown, Tom B., Dandelion Mané, Aurko Roy, Martín Abadi, and Justin Gilmer. 2017. "Adversarial patch." *arXiv preprint arXiv:1712.09665*. https://arxiv.org/abs/1712.09665.

Bughin, Jacques, Jeongmin Seong, James Manyika, Michael Chui, and Raoul Joshi. 2018. "Notes from the frontier: Modeling the impact of AI on the world economy." McKinsey and Co. September 2018. https://www.mckinsey.com/featured-insights/artificial-intelligence/notes-from-the-frontier-modeling-the-impact-of-ai-on-the-world-economy.

Buolamwini, Joy, and Timnit Gebru. 2018. "Gender shades: Intersectional accuracy disparities in commercial gender classification." In *Conference on Fairness, Accountability and Transparency, 2018*. 77–91. http://proceed ings.mlr.press/v81/buolamwini18a.html.

Burns, Janet. 2017. "Finally, the perfect app for superfans, stalkers, and serial killers." *Forbes*. June 23, 2017. https://www.forbes.com/sites/janetw burns/2017/06/23/finally-the-perfect-dating-app-for-superfans-stalkers -and-serial-killers/#4d2b54c9f166.

Bushnell, Mona. 2018. "AI faceoff: Siri vs. Cortana vs. Google Assistant vs. Alexa." *Business News Daily*. June 29, 2018. https://www.businessnews daily.com/10315-siri-cortana-google-assistant-amazon-alexa-face-off .html.

Cable Car Museum, undated. "The Brakes." http://www.cablecarmuseum.org /the-brakes.html. Accessed by the authors, December 29, 2018.

Callahan, John. 2019. "What is Google Duplex, and how do you use it?" *Android Authority*, March 3, 2019. https://www.androidauthority.com /what-is-google-duplex-869476/.

Campolo, Alex, Madelyn Sanfilippo, Meredith Whittaker, and Kate Crawford. 2017. *AI Now 2017 Report*. https://ainowinstitute.org/AI_Now_2017 _Report.pdf.

Canales, Katie. 2018. "A couple says that Amazon's Alexa recorded a private conversation and randomly sent it to a friend." *Business Insider*. May 24, 2018. http://www.businessinsider.com/amazon-alexa-records-private -conversation-2018-5.

Canziani, Alfredo, Eugenio Culurciello, and Adam Paszke. 2017. "Evaluation of neural network architectures for embedded systems." In *IEEE International Symposium on Circuits and Systems (ISCAS), 2017*. 1–4. https:// ieeexplore.ieee.org/abstract/document/8050276/.

Carey, Susan. 1985. *Conceptual Change in Childhood*. Cambridge, MA: MIT Press.

Carmichael, Leonard, H. P. Hogan, and A. A. Walter. 1932. "An experimental study of the effect of language on the reproduction of visually perceived form." *Journal of Experimental Psychology* 15(1): 73. http://dx.doi .org/10.1037/h0072671.

Chaplot, Devendra Singh, Guillaume Lample, Kanthashree Mysore Sathyendra, and Ruslan Salakhutdinov. 2016. "Transfer deep reinforcement learning in 3d environments: An empirical study." In *NIPS Deep Reinforcement Learning Workshop*. http://www.cs.cmu.edu/~rsalakhu/papers /DeepRL_Transfer.pdf

Chintala, Soumith, and Yann LeCun, 2016. "A path to unsupervised learning through adversarial networks." Facebook AI Research blog, June 20, 2016. https://code.fb.com/ml-applications/a-path-to-unsupervised-learning-through-adversarial-networks/.

Chokshi, Niraj. 2018. "Amazon knows why Alexa was laughing at its customers." *New York Times*, March 8, 2018. https://www.nytimes.com/2018/03/08/business/alexa-laugh-amazon-echo.html.

Chomsky, Noam. 1959. "A review of B. F. Skinner's *Verbal Behavior*." *Language* 35(1): 26–58. http://doi.org/10.2307/411334.

Chu-Carroll, Jennifer, James Fan, B. K. Boguraev, David Carmel, Dafna Sheinwald, and Chris Welty. 2012. "Finding needles in the haystack: Search and candidate generation" *IBM Journal of Research and Development* 56(3–4): 6:1–6:12. https://doi.org/10.1147/JRD.2012.2186682.

CNBC. 2018. *Boston Dynamics' Atlas Robot Can Now Do Parkour.* Video. https://www.youtube.com/watch?v=hSjKoEva5bg.

Coldewey, Devin. 2018. "Judge says 'literal but nonsensical' Google translation isn't consent for police search." *TechCrunch*, June 15, 2018. https://techcrunch.com/2018/06/15/judge-says-literal-but-nonsensical-google-translation-isnt-consent-for-police-search/.

Collins, Allan M., and M. Ross Quillian. 1969. "Retrieval time from semantic memory." *Journal of Verbal Learning and Verbal Behavior* 8(2): 240–247. https://doi.org/10.1016/S0022-5371(69)80069-1.

Collins, Harry. 2018. *Artifictional Intelligence: Against Humanity's Surrender to Computers*. New York: Wiley.

Conesa, Jordi, Veda C. Storey, and Vijayan Sugumaran. 2010. "Usability of upper level ontologies: The case of ResearchCyc." *Data & Knowledge Engineering* 69(4): 343–356. https://doi.org/10.1016/j.datak.2009.08.002.

Conneau, Alexis, German Kruszewski, Guillaume Lample, Loïc Barrault, and Marco Baroni. 2018. "What you can cram into a single vector: Probing sentence embeddings for linguistic properties." *arXiv preprint arXiv:1805.01070* https://arxiv.org/pdf/1805.01070.pdf.

Corbett, Erin, and Jonathan Vanian. 2018. "Microsoft improves biased facial recognition technology." *Fortune*. June 27, 2018. http://fortune.com/2018/06/27/microsoft-biased-facial-recognition/.

Crick, Francis. 1989. "The recent excitement about neural networks." *Nature* 337(6203): 129–132. https://doi.org/10.1038/337129a0.

Cuthbertson, Anthony. 2018. "Robots can now read better than humans, putting millions of jobs at risk." *Newsweek*. January 15, 2018. https://www

.newsweek.com/robots-can-now-read-better-humans-putting-millions-jobs-risk-781393.

Damiani, Jesse. 2018. "Tesla Model S on Autopilot crashes into parked police vehicle in Laguna Beach." *Forbes*. May 30, 2018. https://www.forbes.com/sites/jessedamiani/2018/05/30/tesla-model-s-on-autopilot-crashes-into-parked-police-vehicle-in-laguna-beach/#7c5d245d6f59.

Darwiche, Adnan. 2018. "Human-level intelligence or animal-like abilities?" *Communications of the ACM* 61(10): 56–67. https://cacm.acm.org/magazines/2018/10/231373-human-level-intelligence-or-animal-like-abilities/fulltext.

Dastin, Jeffrey. 2018. "Amazon scraps secret AI recruiting tool that showed bias against women." Reuters. October 10, 2018. https://www.reuters.com/article/amazoncom-jobs-automation/rpt-insight-amazon-scraps-secret-ai-recruiting-tool-that-showed-bias-against-women-idUSL2N1WP1RO.

Davies, Alex. 2017. "Waymo has taken the human out of its self-driving cars." *WIRED*. November 7, 2017. https://www.wired.com/story/waymo-google-arizona-phoenix-driverless-self-driving-cars/.

Davies, Alex. 2018. "Waymo's so-called Robo-Taxi launch reveals a brutal truth." *WIRED*. December 5, 2018. https://www.wired.com/story/waymo-self-driving-taxi-service-launch-chandler-arizona/.

Davis, Ernest. 1990. *Representations of Commonsense Knowledge*. San Mateo, CA: Morgan Kaufmann.

Davis, Ernest. 2016a. "How to write science questions that are easy for people and hard for computers." *AI Magazine* 37(1): 13–22.

Davis, Ernest. 2016b. "The tragic tale of Tay the Chatbot." *AI Matters* 2(4). https://cs.nyu.edu/faculty/davise/Verses/Tay.html.

Davis, Ernest. 2017. "Logical formalizations of commonsense reasoning." *Journal of Artificial Intelligence Research* 59: 651–723. https://jair.org/index.php/jair/article/view/11076.

Davis, Ernest, and Gary Marcus. 2015. "Commonsense reasoning and commonsense knowledge in artificial intelligence." *Communications of the ACM* 58(9): 92–105.

Davis, Ernest, and Gary Marcus. 2016. "The scope and limits of simulation in automated reasoning." *Artificial Intelligence* 233: 60–72. http://dx.doi.org/10.1016/j.artint.2015.12.003.

Davis, Randall, and Douglas Lenat. 1982. *Knowledge-Based Systems in Artificial Intelligence*. New York: McGraw-Hill.

Deerwester, Scott, Susan T. Dumais, George W. Furnas, Thomas K. Landauer, and Richard Harshman. 1990. "Indexing by latent semantic analysis." *Journal of the American Society for Information Science* 41(6): 391–407. https://doi.org/10.1002/(SICI)1097-4571(199009)41:6<391::AID-ASI1>3.0.CO;2-9.

Deng, Jia, Wei Dong, Richard Socher, Li-Jia Li, Kai Li, and Li Fei-Fei. "Imagenet: A large-scale hierarchical image database." *IEEE Conference on Computer Vision and Pattern Recognition, 2009.* 248–255. https://doi.org/10.1109/CVPR.2009.5206848.

Dennett, Daniel. 1978. *Brainstorms: Philosophical Essays on Mind and Psychology.* Cambridge, MA: MIT Press.

Devlin, Hannah. 2015. "Google a step closer to developing machines with human-like intelligence." *The Guardian.* May 21, 2015. https://www.theguardian.com/science/2015/may/21/google-a-step-closer-to-developing-machines-with-human-like-intelligence.

Diamandis, Peter, and Steven Kotler. 2012. *Abundance: The Future Is Better Than You Think.* New York: Free Press.

Domingos, Pedro. 2015. *The Master Algorithm: How the Quest for the Ultimate Learning Machine Will Remake Our World.* New York: Basic Books.

D'Orazio, Dante. 2014. "Elon Musk says artificial intelligence is 'potentially more dangerous than nukes.'" *The Verge.* August 3, 2014. https://www.theverge.com/2014/8/3/5965099/elon-musk-compares-artificial-intelligence-to-nukes.

Dreyfus, Hubert. 1979. *What Computers Can't Do: The Limits of Artificial Intelligence.* Rev. ed. New York: Harper and Row.

Dreyfuss, Emily. 2018. "A bot panic hits Amazon's mechanical Turk." *WIRED.* August 17, 2018. https://www.wired.com/story/amazon-mechanical-turk-bot-panic/.

Dyer, Michael. 1983. *In-Depth Understanding: A Computer Model of Integrated Processing for Narrative Comprehension.* Cambridge, MA: MIT Press.

Estava, Andre, Brett Kuprel, Roberto A. Novoa, Justin Ko, Susan M. Swetter, Helen M. Blau, and Sebastian Thrun. 2017. "Dermatologist-level classification of skin cancer with deep neural networks." *Nature* 542(7639): 115–118.

*The Economist.* 2018. "AI, radiology, and the future of work." June 7, 2018. https://www.economist.com/leaders/2018/06/07/ai-radiology-and-the-future-of-work.

Eubanks, Virginia. 2018. *Automating Inequality: How High-Tech Tools Profile, Police, and Punish the Poor.* New York: St. Martin's Press.

Evans, Jonathan St. B. T. 2012. "Dual process theories of deductive reasoning: Facts and fallacies." In *The Oxford Handbook of Thinking and Reasoning*, 115–133. Oxford: Oxford University Press.

Evarts, Eric C. 2016. "Why Tesla's Autopilot isn't really autopilot." *U.S. News and World Report Best Cars*. August 11, 2016. https://cars.usnews.com/cars-trucks/best-cars-blog/2016/08/why-teslas-autopilot-isnt-really-autopilot.

Evtimov, Ivan, Kevin Eykholt, Earlence Fernandes, Tadayoshi Kohno, Bo Li, Atul Prakash, Amir Rahmati, and Dawn Song. 2017. "Robust physical-world attacks on machine learning models." *arXiv preprint arXiv:1707.08945*. https://arxiv.org/abs/1707.08945.

Fabian. 2018. "Global artificial intelligence landscape." *Medium.com*. May 22, 2018. https://medium.com/@bootstrappingme/global-artificial-intelligence-landscape-including-database-with-3-465-ai-companies-3bf01a175c5d.

Falcon, William. 2018. "The new Burning Man—the AI conference that sold out in 12 minutes." *Forbes*. September 5, 2018. https://www.forbes.com/sites/williamfalcon/2018/09/05/the-new-burning-man-the-ai-conference-that-sold-out-in-12-minutes/#38467b847a96.

Felleman, Daniel J., and D. C. van Essen. 1991. "Distributed hierarchical processing in the primate cerebral cortex." *Cerebral Cortex* 1(1): 1–47. https://doi.org/10.1093/cercor/1.1.1-a.

Fernandez, Ernie. 2016. "How cognitive systems will shape the future of health and wellness." *IBM Healthcare and Life Sciences Industry Blog*. November 16, 2016. https://www.ibm.com/blogs/insights-on-business/healthcare/cognitive-systems-shape-health-wellness/.

Ferrucci, David, Eric Brown, Jennifer Chu-Carroll, James Fan, David Gondek, Aditya A. Kalyanpur, Adam Lally, et al. 2010. "Building Watson: An overview of the DeepQA project." *AI Magazine* 31(3): 59–79. https://www.aaai.org/ojs/index.php/aimagazine/article/view/2303.

Firestone, Chaz, and Brian J. Scholl. 2016. "Cognition does not affect perception: Evaluating the evidence for 'top-down' effects." *Behavioral and Brain Sciences* 39, e229. https://doi.org/10.1017/S0140525X15000965.

Ford, Martin. 2018. *Architects of Intelligence: The Truth About AI from the People Building It*. Birmingham, UK: Packt Publishing.

Fukushima, Kunihiko, and Sei Miyake. 1982. "Neocognitron: A self-organizing neural network model for a mechanism of visual pattern recognition." In *Competition and Cooperation in Neural Nets: Proceedings of the U.S.–*

*Japan Joint Seminar,* 267–285. Berlin, Heidelberg: Springer. https://doi.org/10.1007/978-3-642-46466-9_18.

Fung, Brian. 2017. "The driver who died in a Tesla crash using Autopilot ignored at least 7 safety warnings." *Washington Post.* June 20, 2017. https://www.washingtonpost.com/news/the-switch/wp/2017/06/20/the-driver-who-died-in-a-tesla-crash-using-autopilot-ignored-7-safety-warnings/.

Future of Life Institute. 2015. "Autonomous weapons: An open letter from AI & robotics researchers." https://futureoflife.org/open-letter-autonomous-weapons/.

Gardner, Howard. 1983. *Frames of Mind: The Theory of Multiple Intelligences.* New York: Basic Books.

Garrahan, Matthew. 2017. "Google and Facebook dominance forecast to rise." *Financial Times.* December 3, 2017. https://www.ft.com/content/cf362186-d840-11e7-a039-c64b1c09b482.

Gatys, Leon A., Alexander S. Ecker, and Matthias Bethge. 2016. "Image style transfer using convolutional neural networks." In *Proceedings of the IEEE Conference on Computer Vision and Pattern Recognition,* 2414–2423. https://www.cv-foundation.org/openaccess/content_cvpr_2016/html/Gatys_Image_Style_Transfer_CVPR_2016_paper.html.

Geirhos, Robert, Carlos R. M. Temme, Jonas Rauber, Heiko H. Schütt, Matthias Bethge, and Felix A. Wichmann. 2018. "Generalisation in humans and deep neural networks." In *Advances in Neural Information Processing Systems,* 7549–7561. http://papers.nips.cc/paper/7982-generalisation-in-humans-and-deep-neural-networks.

Gelman, Rochel, and Renee Baillargeon. 1983. "Review of some Piagetian concepts." In *Handbook of Child Psychology: Formerly Carmichael's Manual of Child Psychology,* edited by Paul H. Mussen. New York: Wiley.

Geman, Stuart, Elie Bienenstock, and René Doursat. 1992. "Neural networks and the bias/variance dilemma." *Neural Computation* 4(1): 1–58. https://doi.org/10.1162/neco.1992.4.1.1.

Gershgorn, Dave. 2017. "The data that transformed AI research—and possibly the world." *Quartz.* July 26, 2017. https://qz.com/1034972/the-data-that-changed-the-direction-of-ai-research-and-possibly-the-world/.

Gibbs, Samuel. 2014. "Google buys UK artificial intelligence startup Deepmind for £400m." *The Guardian.* January 27, 2014. https://www.theguardian.com/technology/2014/jan/27/google-acquires-uk-artificial-intelligence-startup-deepmind.

Gibbs, Samuel. 2018. "SpotMini: Headless robotic dog to go on sale in 2019." *The Guardian*. May 14, 2018. https://www.theguardian.com/technology/2018/may/14/spotmini-robotic-dog-sale-2019-former-google-boston-dynamics.

Glaser, April. 2018. "The robot dog that can open a door is even more impressive than it looks." *Slate*. February 13, 2018. https://slate.com/technology/2018/02/the-robot-dog-that-can-open-a-door-is-even-more-impressive-than-it-looks.html.

Glasser, Matthew, et al. 2016. "A multi-modal parcellation of human cerebral cortex." *Nature* 536: 171–178. https://doi.org/10.1038/nature18933.

Glorot, Xavier, Antoine Bordes, and Yoshua Bengio. 2011. "Deep sparse rectifier neural networks." In *Proceedings of the Fourteenth International Conference on Artificial Intelligence and Statistics,* 315–323. http://proceedings.mlr.press/v15/glorot11a/glorot11a.pdf.

Goode, Lauren. 2018. "Google CEO Sundar Pichai compares impact of AI to electricity and fire." *The Verge*. Jan. 19, 2018. https://www.theverge.com/2018/1/19/16911354/google-ceo-sundar-pichai-ai-artificial-intelligence-fire-electricity-jobs-cancer.

Goodfellow, Ian, Yoshua Bengio, and Aaron Courville. 2015. *Deep Learning*. Cambridge, MA: MIT Press.

Greenberg, Andy. 2015. "Hackers remotely kill a Jeep on the highway—with me in it." *WIRED*. July 21, 2015. https://www.wired.com/2015/07/hackers-remotely-kill-jeep-highway/.

Greenberg, Andy. 2017. "Watch a 10-year-old's face unlock his mom's iPhone X." *WIRED*. November 14, 2017. https://www.wired.com/story/10-year-old-face-id-unlocks-mothers-iphone-x/.

Greg. 2018. "Dog's final judgement: Weird Google Translate glitch delivers an apocalyptic message." *Daily Grail*, July 16, 2018. https://www.dailygrail.com/2018/07/dogs-final-judgement-weird-google-translate-glitch-delivers-an-apocalyptic-message/.

Gunning, David. 2017. "Explainable artificial intelligence (xai)." *Defense Advanced Research Projects Agency (DARPA)*. https://www.darpa.mil/attachments/XAIProgramUpdate.pdf.

Hall, Phil. 2018. "Luminar's smart Sky Enhancer filter does the dodging and burning for you." *Techradar: The Source for Tech Buying Advice*. November 2, 2018. https://www.techradar.com/news/luminars-smart-sky-enhancer-filter-does-the-dodging-and-burning-for-you.

Harford, Tim. 2018. "What we get wrong about technology." *Financial Times*. July 7, 2018. https://www.ft.com/content/32c31874-610b-11e7-8814-0ac7eb84e5f1.

Harridy, Rich. 2018. "Boston Dynamics Atlas robot can now chase you through the woods." *New Atlas*. May 10, 2018. https://newatlas.com/boston-dynamics-atlas-running/54573/.

Harwell, Drew. 2018. "Elon Musk said a Tesla could drive itself across the country by 2018. One just crashed backing out of a garage." *Washington Post*. September 13, 2018. https://www.washingtonpost.com/technology/2018/09/13/elon-musk-said-tesla-could-drive-itself-across-country-by-one-just-crashed-backing-out-garage/.

Harwell, Drew, and Craig Timberg. 2019. "YouTube recommended a Russian media site thousands of times for analysis of Mueller's report, a watchdog group says." *The Washington Post*, April 26, 2019. https://www.washingtonpost.com/technology/2019/04/26/youtube-recommended-russian-media-site-above-all-others-analysis-mueller-report-watchdog-group-says/?utm_term=.39b7bcf0c8a4.

Havasi, Catherine, Robert Speer, James Pustejovsky, and Henry Lieberman. 2009. "Digital intuition: Applying common sense using dimensionality reduction." *IEEE Intelligent systems* 24(4): 24–35. https://doi.org/10.1109/MIS.2009.72.

Hawkins, Andrew. 2018. "Elon Musk still doesn't think LIDAR is necessary for fully driverless cars." *The Verge*. February 7, 2018. https://www.theverge.com/2018/2/7/16988628/elon-musk-lidar-self-driving-car-tesla.

Hawkins, Jeff, and Sandra Blakeslee. 2004. *On Intelligence: How a New Understanding of the Brain Will Lead to the Creation of Truly Intelligent Machines*. New York: Times Books.

Hayes, Gavin. 2018. "Search 'idiot,' get Trump: How activists are manipulating Google Images." *The Guardian*. July 17, 2018. https://www.theguardian.com/us-news/2018/jul/17/trump-idiot-google-images-search.

Hayes, Patrick, and Kenneth Ford. 1995. "Turing test considered harmful." *Intl. Joint Conf. on Artificial Intelligence:* 972–977. https://www.researchgate.net/profile/Kenneth_Ford/publication/220813820_Turing_Test_Considered_Harmful/links/09e4150d1dc67df32c000000.pdf.

Hazelwood, Kim, et al. 2017. "Applied machine learning at Facebook: A data center infrastructure perspective." https://research.fb.com/wp-content/uploads/2017/12/hpca-2018-facebook.pdf.

He, Kaiming, Xiangyu Zhang, Shaoqing Ren, and Jian Sun. 2016. "Deep residual learning for image recognition." In *Proceedings of the IEEE Conference on Computer Vision and Pattern Recognition:* 770–778. https://www.cv-foundation.org/openaccess/content_cvpr_2016/html/He_Deep_Residual_Learning_CVPR_2016_paper.html.

He, Mingming, Dongdong Chen, Jing Liao, Pedro V. Sander, and Lu Yuan. 2018. "Deep exemplar-based colorization." *ACM Transactions on Graphics* 37(4): Article 47. https://doi.org/10.1145/3197517.3201365.

He, Xiaodong, and Li Deng. 2017. "Deep learning for image-to-text generation: A technical overview." *IEEE Signal Processing Magazine* 34(6): 109–116. https://doi.org/10.1109/MSP.2017.2741510.

Heath, Nick. 2018. "Google DeepMind founder Demis Hassabis: Three truths about AI." *Tech Republic*. September 24, 2018. https://www.techrepublic.com/article/google-deepmind-founder-demis-hassabis-three-truths-about-ai/.

Herculano-Houzel, Suzana. 2016. *The Human Advantage: A New Understanding of How Our Brains Became Remarkable*. Cambridge, MA: MIT Press.

Herman, Arthur. 2018. "China's brave new world of AI." *Forbes*. August 30, 2018. https://www.forbes.com/sites/arthurherman/2018/08/30/chinas-brave-new-world-of-ai/#1051418628e9.

Hermann, Karl Moritz, Felix Hill, Simon Green, Fumin Wang, Ryan Faulkner, Hubert Soyer, David Szepesvari, et al. 2017. "Grounded language learning in a simulated 3D world." *arXiv preprint arXiv:1706.06551*. https://arxiv.org/abs/1706.06551.

Herper, Matthew. 2017. "M. D. Anderson benches IBM Watson in setback for artificial intelligence in medicine." *Forbes*. February 19, 2017. https://www.forbes.com/sites/matthewherper/2017/02/19/md-anderson-benches-ibm-watson-in-setback-for-artificial-intelligence-in-medicine/#319104243774.

Hines, Matt. 2007. "Spammers establishing use of artificial intelligence." *Computer World*. June 1, 2007. https://www.computerworld.com/article/2541475/security0/spammers-establishing-use-of-artificial-intelligence.html.

Hinton, Geoffrey E., Terrence Joseph Sejnowski, and Tomaso A. Poggio, eds. 1999. *Unsupervised Learning: Foundations of Neural Computation*. Cambridge, MA: MIT Press.

Hof, Robert D. 2013. "10 breakthrough technologies: Deep learning." *MIT Technology Review*. https://www.technologyreview.com/s/513696/deep-learning/.

Hoffman, Judy, Dequan Wang, Fisher Yu, and Trevor Darrell. 2016. "FCNs in the wild: Pixel-level adversarial and constraint-based adaptation." *arXiv preprint arXiv:1612.02649*. https://arxiv.org/abs/1612.02649.

Hofstadter, Douglas. 2018. "The shallowness of Google Translate." *The Atlantic*. January 30, 2018. https://www.theatlantic.com/technology/archive/2018/01/the-shallowness-of-google-translate/551570/.

Hornyak, Tim. 2018. "Sony's new dog Aibo barks, does tricks, and charms animal lovers." *CNBC*. April 9, 2018. https://www.cnbc.com/2018/04/09/sonys-new-robot-dog-aibo-barks-does-tricks-and-charms-animal-lovers.html.

Hosseini, Hossein, Baicen Xiao, Mayoore Jaiswal, and Radha Poovendran. 2017. "On the limitation of convolutional neural networks in recognizing negative images." In *2017 16th IEEE International Conference on Machine Learning and Applications (ICMLA)*: 352–358. https://doi.org/10.1109/ICMLA.2017.0-136.

Hua, Sujun, and Zhirong Sun. 2001. "A novel method of protein secondary structure prediction with high segment overlap measure: Support vector machine approach." *Journal of Molecular Biology* 308(2): 397–407. https://doi.org/10.1006/jmbi.2001.4580.

Huang, Sandy, Nicolas Papernot, Ian Goodfellow, Yan Duan, and Pieter Abbeel. 2017. "Adversarial attacks on neural network policies." *arXiv preprint arXiv:1702.02284*. https://arxiv.org/abs/1702.02284.

Huang, Xuedong, James Baker, and Raj Reddy. "A historical perspective of speech recognition." *Communications of the ACM* 57(1): 94–103. https://m-cacm.acm.org/magazines/2014/1/170863-a-historical-perspective-of-speech-recognition/.

Hubel, David H., and Torsten N. Wiesel. 1962. "Receptive fields, binocular interaction and functional architecture in the cat's visual cortex." *Journal of Physiology* 160(1): 106–154. https://doi.org/10.1113/jphysiol.1962.sp006837.

Huff, Darrell. 1954. *How to Lie with Statistics*. New York, W. W. Norton.

IBM Watson Health. 2016. "Five ways cognitive technology can revolutionize healthcare." *Watson Health Perspectives*. October 28, 2016. https://www.ibm.com/blogs/watson-health/5-ways-cognitive-technology-can-help-revolutionize-healthcare/.

IBM Watson Health. Undated. "Welcome to the cognitive era of health." *Watson Health Perspectives*. http://www-07.ibm.com/hk/watson/health/. Accessed by the authors, December 23, 2018.

IEEE Spectrum. 2015. *A Compilation of Robots Falling Down at the DARPA*

*Robotics Challenge*. Video. Posted to YouTube June 6, 2015. https://www.youtube.com/watch?v=g0TaYhjpOfo.

Iizuka, Satoshi, Edgar Simo-Serra, and Hiroshi Ishikawa. 2016. "Let there be color!: Joint end-to-end learning of global and local image priors for automatic image colorization with simultaneous classification." *ACM Transactions on Graphics (TOG)* 35(4): 110. https://dl.acm.org/citation.cfm?id=2925974.

Irpan, Alex. 2018. "Deep reinforcement learning doesn't work yet." *Sorta Insightful* blog. February 14, 2018. https://www.alexirpan.com/2018/02/14/rl-hard.html.

Jeannin, Jean-Baptiste, Khalil Ghorbal, Yanni Kouskoulas, Ryan Gardner, Aurora Schmidt, Erik Zawadzki, and André Platzer. 2015. "A formally verified hybrid system for the next-generation airborne collision avoidance system." In *International Conference on Tools and Algorithms for the Construction and Analysis of Systems:* 21–36. Berlin, Heidelberg: Springer. http://ra.adm.cs.cmu.edu/anon/home/ftp/2014/CMU-CS-14-138.pdf.

Jia, Robin, and Percy Liang. 2017. "Adversarial examples for evaluating reading comprehension systems." *arXiv preprint arXiv:1707.07328*. https://arxiv.org/abs/1707.07328.

Jo, Jason, and Yoshua Bengio. 2017. "Measuring the tendency of CNNs to learn surface statistical regularities." *arXiv preprint arXiv:1711.11561*. https://arxiv.org/abs/1711.11561.

Joachims, T. 2002. *Learning to Classify Text Using Support Vector Machines: Methods, Theory and Algorithms*. Boston: Kluwer Academic Publishers.

Judson, Horace. 1980. *The Eighth Day of Creation: Makers of the Revolution in Biology*. New York: Simon and Schuster.

Jurafsky, Daniel, and James H. Martin. 2009. *Speech and Language Processing*. 2nd ed. Upper Saddle River, NJ: Pearson.

Kahneman, Daniel. 2011. *Thinking, Fast and Slow*. New York: Farrar, Straus, and Giroux.

Kahneman, Daniel, Anne Treisman, and Brian J. Gibbs. 1992. "The reviewing of object files: Object-specific integration of information." *Cognitive Psychology* 24(2): 175–219. https://doi.org/10.1016/0010-0285(92)90007-O.

Kandel, Eric, James Schwartz, and Thomas Jessell. 1991. *Principles of Neural Science*. Norwalk, CT: Appleton & Lange.

Kansky, Ken, Tom Silver, David A. Mély, Mohamed Eldawy, Miguel Lázaro-Gredilla, Xinghua Lou, Nimrod Dorfman, Szymon Sidor, Scott Phoenix, and Dileep George. 2017. "Schema networks: Zero-shot transfer with a generative causal model of intuitive physics." *arXiv preprint arXiv:1706.04317*. https://arxiv.org/abs/1706.04317.

Kant, Immanuel. 1751/1998. *Critique of Pure Reason*. Trans. Paul Guyer and Allen Wood. Cambridge: Cambridge University Press.

Karmon, Danny, Daniel Zoran, and Yoav Goldberg. 2018. "LaVAN: Localized and Visible Adversarial Noise." *arXiv preprint arXiv:1801.02608*. https://arxiv.org/abs/1801.02608.

Kastranakes, Jacob. 2017. "GPS will be accurate within one foot in some phones next year." *The Verge*. September 25, 2017. https://www.theverge.com/circuitbreaker/2017/9/25/16362296/gps-accuracy-improving-one-foot-broadcom.

Keil, Frank C. 1992. *Concepts, Kinds, and Cognitive Development*. Cambridge, MA: MIT Press.

Kim, Sangbae, Cecilia Laschi, and Barry Trimmer. 2013. "Soft robotics: A bioinspired evolution in robotics." *Trends in Biotechnology* 31(5): 287–294. https://doi.org/10.1016/j.tibtech.2013.03.002.

Kintsch, Walter, and Teun A. Van Dijk. 1978. "Toward a model of text comprehension and production." *Psychological Review* 85(5): 363–394.

Kinzler, Katherine D., and Elizabeth S. Spelke. 2007. "Core systems in human cognition." *Progress in Brain Research* 164: 257–264. https://doi.org/10.1016/S0079-6123(07)64014-X.

Kissinger, Henry. 2018. "The End of the Enlightenment." *The Atlantic*, June 2018. https://www.theatlantic.com/magazine/archive/2018/06/henry-kissinger-ai-could-mean-the-end-of-human-history/559124/.

Koehn, Philipp, and Rebecca Knowles. 2017. "Six challenges for neural machine translation." *Proceedings of the First Workshop on Neural Machine Translation*. http://www.aclweb.org/anthology/W/W17/W17-3204.pdf.

Krakovna, Victoria. 2018. "Specification gaming examples in AI." Blog post. April 2, 2018. https://vkrakovna.wordpress.com/2018/04/02/specification-gaming-examples-in-ai/.

Krizhevsky, A., I. Sutskever, and G. E. Hinton. 2012. "ImageNet classification with deep convolutional neural networks." In *Advances in Neural Information Processing Systems:* 1097–1105. http://papers.nips.cc/paper/4824-imagenet-classification-with-deep-convolutional-neural-networks.pdf.

Kurzweil, Ray. 2002. "Response to Mitchell Kapor's "Why I Think I Will Win." *Kurzweil Accelerating Intelligence Essays*. http://www.kurzweilai.net/response-to-mitchell-kapor-s-why-i-think-i-will-win.

Kurzweil, Ray. 2013. *How to Create a Mind: The Secret of Human Thought Revealed.* New York: Viking.

Kurzweil, Ray, and Rachel Bernstein. 2018. "Introducing semantic experiences with Semantris and Talk to Books." *Google AI Blog*. April 13, 2018. https://ai.googleblog.com/2018/04/introducing-semantic-experiences-with.html.

Lancaster, Luke. 2016. "Elon Musk's OpenAI is working on a robot butler." *CNet*. June 22, 2016. https://www.cnet.com/news/elon-musks-openai-is-working-on-a-robot-butler/.

Lardieri, Alexa. 2018. "Drones deliver life-saving blood to remote African regions." *US News & World Report*. January 2, 2018.

Lashbrook, Angela. 2018. "AI-driven dermatology could leave dark-skinned patients behind." *The Atlantic*. August 16, 2018. https://www.theatlantic.com/health/archive/2018/08/machine-learning-dermatology-skin-color/567619/.

LaValle, Stephen M. 2006. *Planning Algorithms.* Cambridge: Cambridge University Press.

Leben, Derek. 2018. *Ethics for Robots: How to Design a Moral Algorithm.* Milton Park, UK: Routledge.

Lecoutre, Adrian, Benjamin Negrevergne, and Florian Yger. 2017. "Recognizing art style automatically in painting with deep learning." *Proceedings of the Ninth Asian Conference on Machine Learning, PMLR* 77: 327–342. http://proceedings.mlr.press/v77/lecoutre17a.html.

LeCun, Yann. 2018. "Research and projects." http://yann.lecun.com/ex/research/index.html. Accessed by the authors, September 6, 2018.

LeCun, Yann, Bernhard Boser, John S. Denker, Donnie Henderson, Richard E. Howard, Wayne Hubbard, and Lawrence D. Jackel. 1989. "Backpropagation applied to handwritten zip code recognition." *Neural Computation* 1(4): 541–551. https://www.mitpressjournals.org/doi/abs/10.1162/neco.1989.1.4.541.

LeCun, Yann, and Yoshua Bengio. 1995. "Convolutional networks for images, speech, and time series." In *The Handbook of Brain Theory and Neural Networks,* edited by Michael Arbib. Cambridge, MA: MIT Press. https://www.researchgate.net/profile/Yann_Lecun/publication/2453996_Convolutional_Networks_for_Images_Speech_and_Time-Series/links/0deec519dfa2325502000000.pdf.

LeCun, Yann, Yoshua Bengio, and Geoffrey Hinton. 2015. "Deep learning." *Nature* 521(7553): 436–444. https://doi.org/10.1038/nature14539.

Lenat, Douglas B., Mayank Prakash, and Mary Shepherd. 1985. "CYC: Using common sense knowledge to overcome brittleness and knowledge acquisition bottlenecks." *AI Magazine* 6(4): 65–85. https://doi.org/10.1609/aimag.v6i4.510.

Lenat, Douglas B., and R. V. Guha. 1990. *Building Large Knowledge-Based Systems: Representation and Inference in the CYC Project*. Boston: Addison-Wesley.

Levesque, Hector. 2017. *Common Sense, the Turing Test, and the Quest for Real AI*. Cambridge, MA: MIT Press.

Levesque, Hector, Ernest Davis, and Leora Morgenstern. 2012. "The Winograd Schema challenge." *Principles of Knowledge Representation and Reasoning, 2012*. http://www.aaai.org/ocs/index.php/KR/KR12/paper/download/4492/4924.

Leviathan, Yaniv. 2018. "Google Duplex: An AI system for accomplishing real-world tasks over the phone." *Google AI Blog*. May 8, 2018. https://ai.googleblog.com/2018/05/duplex-ai-system-for-natural-conversation.html.

Levin, Alan, and Harry Suhartono. 2019. "Pilot who hitched a ride saved Lion Air 737 day before deadly crash." *Bloomberg*. March 19, 2019. https://www.bloomberg.com/news/articles/2019-03-19/how-an-extra-man-in-cockpit-saved-a-737-max-that-later-crashed.

Levin, Sam, and Nicky Woolf. 2016. "Tesla driver killed while using Autopilot was watching Harry Potter, witness says." *The Guardian*. July 3, 2016. https://www.theguardian.com/technology/2016/jul/01/tesla-driver-killed-autopilot-self-driving-car-harry-potter.

Levine, Alexandra S. 2017. "New York today: An Ella Fitzgerald centenary." *New York Times*. April 25, 2017. https://www.nytimes.com/2017/04/25/nyregion/new-york-today-ella-fitzgerald-100th-birthday-centennial.html.

Levy, Omer. In preparation. "Word representations." In *The Oxford Handbook of Computational Linguistics*. 2nd ed. Edited by Ruslan Mitkov. Oxford: Oxford University Press.

Lewis, Dan. 2016. "They Blue It." Now I Know website. March 3, 2016. Accessed by authors, December 25, 2018. http://nowiknow.com/they-blue-it/.

Lewis-Krauss, Gideon. 2016. "The great AI awakening." *New York Times Magazine*. December 14, 2016. https://www.nytimes.com/2016/12/14/magazine/the-great-ai-awakening.html.

Liao, Shannon. 2018. "Chinese facial recognition system mistakes a face on a bus for a jaywalker." *The Verge*. November 22, 2018. https://www.theverge.com/2018/11/22/18107885/china-facial-recognition-mistaken-jaywalker.

Lifschitz, Vladimir, Leora Morgenstern, and David Plaisted. 2008. "Knowledge representation and classical logic." In *Handbook of Knowledge Representation,* edited by Frank van Harmelen, Vladimir Lifschitz, and Bruce Porter, 3–88. Amsterdam: Elsevier.

Lin, Patrick, Keith Abney, and George Bekey, eds. 2012. *Robot Ethics: The Ethical and Social Implications of Robotics.* Cambridge, MA: MIT Press.

Linden, Derek S. 2002. "Antenna design using genetic algorithms." In *Proceedings of the 4th Annual Conference on Genetic and Evolutionary Computation,* 1133–1140. https://dl.acm.org/citation.cfm?id=2955690.

Linn, Alison. 2018. "Microsoft creates AI that can read a document and answer questions about it as well as a person." *Microsoft AI Blog.* January 15, 2018. https://blogs.microsoft.com/ai/microsoft-creates-ai-can-read-document-answer-questions-well-person/.

Lippert, John, Bryan Gruley, Kae Inoue, and Gabrielle Coppola. 2018. "Toyota's vision of autonomous cars is not exactly driverless." *Bloomberg Businessweek.* September 19, 2018. https://www.bloomberg.com/news/features/2018-09-19/toyota-s-vision-of-autonomous-cars-is-not-exactly-driverless.

Lipton, Zachary. 2016. "The mythos of model interpretability." *arXiv preprint arXiv:1606.03490.* https://arxiv.org/abs/1606.03490.

Lupyan, Gary, and Andy Clark. 2015. "Words and the world: Predictive coding and the language-perception-cognition interface." *Current Directions in Psychological Science* 24(4): 279–284. https://doi.org/10.1177/0963721415570732.

Lynch, Kevin, and Frank Park. 2017. *Modern Robotics: Mechanics, Planning, and Control.* Cambridge: Cambridge University Press.

Mahairas, Ari, and Peter J. Beshar. 2018. "A Perfect Target for Cybercriminals," *New York Times.* November 19, 2018. https://www.nytimes.com/2018/11/19/opinion/water-security-vulnerability-hacking.html.

Manning, Christopher, and Hinrich Schütze. 1999. *Foundations of Statistical Natural Language Processing.* Cambridge, MA: MIT Press.

Manning, Christopher, Prabhakar Raghavan, and Hinrich Schütze. 2008. *Introduction to Information Retrieval.* Cambridge: Cambridge University Press.

Marcus, Gary. 2001. *The Algebraic Mind: Integrating Connectionism and Cognitive Science.* Cambridge, MA: MIT Press.

Marcus, Gary. 2004. *The Birth of the Mind: How a Tiny Number of Genes Creates the Complexities of Human Thought.* New York: Basic Books.

Marcus, Gary. 2008. *Kluge: The Haphazard Construction of the Human Mind*. Boston: Houghton Mifflin.

Marcus, Gary. 2012a. "Moral machines." *The New Yorker.* November 24, 2012. https://www.newyorker.com/news/news-desk/moral-machines.

Marcus, Gary. 2012b. "Is deep learning a revolution in artificial intelligence?" *The New Yorker.* November 25, 2012. https://www.newyorker.com/news/news-desk/is-deep-learning-a-revolution-in-artificial-intelligence.

Marcus, Gary. 2018a. "Deep learning: A critical appraisal." *arXiv preprint arXiv:1801.00631*. https://arxiv.org/abs/1801.00631.

Marcus, Gary. 2018b. "Innateness, AlphaZero, and artificial intelligence." *arXiv preprint arXiv:1801.05667*. https://arxiv.org/abs/1801.05667.

Marcus, Gary, and Ernest Davis. 2013. "A grand unified theory of everything." *The New Yorker*. May 6, 2013. https://www.newyorker.com/tech/elements/a-grand-unified-theory-of-everything.

Marcus, Gary, and Ernest Davis. 2018. "No, AI won't solve the fake news problem." *New York Times*. October 20, 2018. https://www.nytimes.com/2018/10/20/opinion/sunday/ai-fake-news-disinformation-campaigns.html.

Marcus, Gary, and Jeremy Freeman. 2015. *The Future of the Brain: Essays by the World's Leading Neuroscientists*. Princeton, NJ: Princeton University Press.

Marcus, Gary, Steven Pinker, Michael Ullman, Michelle Hollander, T. John Rosen, Fei Xu, and Harald Clahsen. 1992. "Overregularization in language acquisition." *Monographs of the Society for Research in Child Development* 57(4): 1–178.

Marcus, Gary, Francesca Rossi, and Manuela Veloso. 2016. *Beyond the Turing Test (AI Magazine Special Issue). AI Magazine* 37(1).

Marshall, Aarian. 2017. "After peak hype, self-driving cars enter the trough of disillusionment." *WIRED.* December 29, 2017. https://www.wired.com/story/self-driving-cars-challenges/.

Mason, Matthew. 2018. "Toward robotic manipulation." *Annual Review of Control, Robotics, and Autonomous Systems* 1: 1–28. https://doi.org/10.1146/annurev-control-060117-104848.

Matchar, Emily. 2017. "AI plant and animal identification helps us all be citizen scientists." *Smithsonian.com*. June 7, 2017. https://www.smithsonianmag.com/innovation/ai-plant-and-animal-identification-helps-us-all-be-citizen-scientists-180963525/.

Matsakis, Louise. 2018. "To break a hate-speech detection algorithm, try 'love.'" *WIRED*. September 26, 2018. https://www.wired.com/story/break-hate-speech-algorithm-try-love/.

Matuszek, Cynthia, Michael Witbrock, Robert C. Kahlert, John Cabral, David Schneider, Purvesh Shah, and Doug Lenat. 2005. "Searching for common sense: populating Cyc™ from the web." *In Proc, American Association for Artificial Intelligence:* 1430–1435. http://www.aaai.org/Papers/AAAI/2005/AAAI05-227.pdf.

Mazzei, Patricia, Nick Madigan, and Anemona Hartocollis. 2018. "Several dead after walkway collapse in Miami." *New York Times*. March 15, 2018. https://www.nytimes.com/2018/03/15/us/fiu-bridge-collapse.html.

McCarthy, John, Marvin Minsky, Nathaniel Rochester, and Claude Shannon. 1955. "A proposal for the summer research project on artificial intelligence." Reprinted in *Artificial Intelligence Magazine* 27(4): 26. https://doi.org/10.1609/aimag.v27i4.1904.

McCarthy, John. 1959. "Programs with common sense." *Proc. Symposium on Mechanization of Thought Processes I.*

McClain, Dylan Loeb. 2011. "First came the machine that defeated a chess champion." *New York Times*. February 16, 2011. https://www.nytimes.com/2011/02/17/us/17deepblue.html.

McDermott, Drew. 1976. "Artificial intelligence meets natural stupidity." *ACM SIGART Bulletin* (57): 4–9. https://doi.org/10.1145/1045339.1045340.

McFarland, Matt. 2014. "Elon Musk: 'With artificial intelligence we are summoning the demon.'" *Washington Post*, October 24, 2014. https://www.washingtonpost.com/news/innovations/wp/2014/10/24/elon-musk-with-artificial-intelligence-we-are-summoning-the-demon/.

McMillan, Robert. 2013. "Google hires brains that helped supercharge machine learning." *WIRED*. March 13, 2013. https://www.wired.com/2013/03/google-hinton/.

Metz, Cade. 2015. "Facebook's human-powered assistant may just supercharge AI." *WIRED*. August 26, 2015. https://www.wired.com/2015/08/how-facebook-m-works/.

Metz, Cade. 2017. "Tech giants are paying huge salaries for scarce A.I. talent." *New York Times*. October 22, 2017. https://www.nytimes.com/2017/10/22/technology/artificial-intelligence-experts-salaries.html.

Metz, Rachel. 2015. "Facebook AI software learns and answers questions." *MIT Technology Review*. March 26, 2015. https://www.technologyreview.com/s/536201/facebook-ai-software-learns-and-answers-questions/.

Mikolov, Tomas, Ilya Sutskever, Kai Chen, Greg Corrado, and Jeffery Dean.

2013. "Distributed representations of words and phrases and their compositionality." *arXiv preprint arXiv:1310.4546*. https://arxiv.org/abs/1310.4546.

Miller, George A. 1995. "WordNet: A lexical database for English." *Communications of the ACM* 38(11): 39–41. https://doi.org/10.1145/219717.219748.

Minsky, Marvin. 1967. *Computation: Finite and Infinite Machines*. Englewood Cliffs, NJ: Prentice Hall.

Minsky, Marvin. 1986. *Society of Mind*. New York: Simon and Schuster.

Minsky, Marvin, and Seymour Papert. 1969. *Perceptrons: An Introduction to Computational Geometry*. Cambridge, MA: MIT Press.

Mitchell, Tom. 1997. *Machine Learning*. New York: McGraw-Hill.

Mnih, Volodymyr, Koray Kavukcuoglu, David Silver, Andrei A. Rusu, Joel Veness, Marc G. Bellemare, Alex Graves, et al. 2015. "Human-level control through deep reinforcement learning." *Nature* 518(7540): 529–533. https://doi.org/10.1038/nature14236.

Molina, Brett. 2017. "Hawking: AI could be 'worst event in the history of our civilization.'" *USA Today*. November 7, 2017. https://www.usatoday.com/story/tech/talkingtech/2017/11/07/hawking-ai-could-worst-event-history-our-civilization/839298001/.

Mouret, Jean-Baptiste, and Konstantinos Chatzilygeroudis. 2017. "20 years of reality gap: A few thoughts about simulators in evolutionary robotics." In *Proceedings of the Genetic and Evolutionary Computation Conference Companion*, 1121–1124. https://doi.org/10.1145/3067695.3082052.

Mueller, Erik. 2006. *Commonsense Reasoning*. Amsterdam: Elsevier Morgan Kaufmann.

Müller, Andreas, and Sarah Guido. 2016. *Introduction to Machine Learning with Python*. Cambridge, MA: O'Reilly Pubs.

Müller, Martin U. 2018. "Playing doctor with Watson: Medical applications expose current limits of AI." *Spiegel Online*. August 3, 2018. http://www.spiegel.de/international/world/playing-doctor-with-watson-medical-applications-expose-current-limits-of-ai-a-1221543.html.

Murphy, Gregory L., and Douglas L. Medin. 1985. "The role of theories in conceptual coherence." *Psychological Review* 92(3): 289–316. http://doi:10.1037/0033-295X.92.3.289.

Murphy, Kevin. 2012. *Machine Learning: A Probabilistic Perspective*. Cambridge, MA: MIT Press.

Murphy, Tom, VII. 2013. "The first level of Super Mario Bros. is easy with lexicographic orderings and time travel . . . after that it gets a little tricky."

*SIGBOVIK* (April 1, 2013). https://www.cs.cmu.edu/~tom7/mario/mario.pdf.

Nandi, Manojit. 2015. "Faster deep learning with GPUs and Theano." *Domino Data Science Blog.* August 4, 2015. https://blog.dominodatalab.com/gpu-computing-and-deep-learning/.

NCES (National Center for Education Statistics). 2019. "Fast Facts: Race/ethnicity of college faculty." Downloaded April 8, 2019.

*New York Times.* 1958. "Electronic 'brain' teaches itself." July 13, 1958. https://www.nytimes.com/1958/07/13/archives/electronic-brain-teaches-itself.html.

Newell, Allen. 1982. "The knowledge level." *Artificial Intelligence* 18(1): 87–127.

Newton, Casey. 2018. "Facebook is shutting down M, its personal assistant service that combined humans and AI." *The Verge.* January 8, 2018. https://www.theverge.com/2018/1/8/16856654/facebook-m-shutdown-bots-ai.

Ng, Andrew. 2016. "What artificial intelligence can and can't do right now." *Harvard Business Review.* November 9, 2016. https://hbr.org/2016/11/what-artificial-intelligence-can-and-cant-do-right-now.

Ng, Andrew, Daishi Harada, and Stuart Russell. 1999. "Policy invariance under reward transformations: Theory and application to reward shaping." In *Int. Conf. on Machine Learning* 99: 278–287. http://luthuli.cs.uiuc.edu/~daf/courses/games/AIpapers/ng99policy.pdf.

Norouzzadeh, Mohammad Sadegh, Anh Nguyen, Margaret Kosmala, Alexandra Swanson, Meredith S. Palmer, Craig Packer, and Jeff Clune. 2018. "Automatically identifying, counting, and describing wild animals in camera-trap images with deep learning." *Proceedings of the National Academy of Sciences* 115(25): E5716–E5725. https://doi.org/10.1073/pnas.1719367115.

Norvig, Peter. 1986. *Unified Theory of Inference for Text Understanding.* PhD thesis, University of California at Berkeley.

Oh, Kyoung-Su, and Keechul Jung. 2004. "GPU implementation of neural networks." *Pattern Recognition* 37(6): 1311–1314.

O'Neil, Cathy. 2016a. *Weapons of Math Destruction: How Big Data Increases Inequality and Threatens Democracy.* New York: Crown.

O'Neil, Cathy. 2016b. "I'll stop calling algorithms racist when you stop anthropomorphizing AI." *Mathbabe* (blog). April 7, 2016. https://mathbabe.org/2016/04/07/ill-stop-calling-algorithms-racist-when-you-stop-anthropomorphizing-ai/.

O'Neil, Cathy. 2017. "The Era of Blind Faith in Big Data Must End." TED talk. https://www.ted.com/talks/cathy_o_neil_the_era_of_blind_faith_in_big_data_must_end/transcript?language=en.

OpenAI. 2018. "Learning Dexterity." *OpenAI* (blog). July 30, 2018. https://blog.openai.com/learning-dexterity/.

Oremus, Will. 2016. "Facebook thinks it has found the secret to making bots less dumb." *Slate*. June 28, 2016. https://slate.com/technology/2016/06/facebooks-a-i-researchers-are-making-bots-smarter-by-giving-them-memory.html.

O'Rourke, Nancy A., Nicholas C. Weiler, Kristina D. Micheva, and Stephen J. Smith. 2012. "Deep molecular diversity of mammalian synapses: why it matters and how to measure it." *Nature Reviews Neuroscience* 13(6): 365–379. https://doi.org/10.1038/nrn3170.

Ortiz, Charles L., Jr. 2016. "Why we need a physically embodied Turing test and what it might look like." *AI Magazine* 37(1): 55–62.

Padfield, Gareth D. 2008. *Helicopter Flight Dynamics: The Theory and Application of Flying Qualities and Simulation Modelling*. New York: Wiley, 2008.

Page, Lawrence, Sergey Brin, Rajeev Motwani, and Terry Winograd. 1999. "The PageRank citation ranking: Bringing order to the web." Technical Report, Stanford InfoLab. http://ilpubs.stanford.edu:8090/422/.

Parish, Peggy. 1963. *Amelia Bedelia*. New York: Harper and Row.

Paritosh, Praveen, and Gary Marcus. 2016. "Toward a comprehension challenge, using crowdsourcing as a tool." *AI Magazine* 37(1): 23–30.

Parker, Stephanie. 2018. "Robot lawnmowers are killing hedgehogs." *WIRED*. September 26, 2018. https://www.wired.com/story/robot-lawnmowers-are-killing-hedgehogs/.

Pearl, Judea, and Dana Mackenzie. 2018. *The Book of Why: The New Science of Cause and Effect*. New York: Basic Books.

Peng, Tony. 2018. "OpenAI Founder: Short-term AGI is a serious possibility." *Medium.com*. November 13, 2018. https://medium.com/syncedreview/openai-founder-short-term-agi-is-a-serious-possibility-368424f7462f.

Pham, Cherise, 2018. "Computers are getting better than humans at reading." *CNN Business*. January 16, 2018. https://money.cnn.com/2018/01/15/technology/reading-robot-alibaba-microsoft-stanford/index.html.

Piaget, Jean. 1928. *The Child's Conception of the World*. London: Routledge and Kegan Paul.

Piantadosi, Steven T. 2014. "Zipf's word frequency law in natural language: A critical review and future directions." *Psychonomic Bulletin & Review* 21(5): 1112–1130. https://www.ncbi.nlm.nih.gov/pmc/articles/PMC4176592.

Piantadosi, Steven T., Harry Tily, and Edward Gibson. 2012. "The communicative function of ambiguity in language." *Cognition* 122(3): 280–291. https://doi.org/10.1016/j.cognition.2011.10.004.

Ping, David, Bing Xiang, Patrick Ng, Ramesh Nallapati, Saswata Chakravarty, and Cheng Tang. 2018. "Introduction to Amazon SageMaker Object2Vec." *AWS Machine Learning* (blog). https://aws.amazon.com/blogs/machine-learning/introduction-to-amazon-sagemaker-object2vec/.

Pinker, Steven. 1994. *The Language Instinct: How the Mind Creates Language.* New York: William Morrow.

Pinker, Steven. 1997. *How the Mind Works.* New York: W. W. Norton.

Pinker, Steven. 1999. *Words and Rules: The Ingredients of Language.* New York: Basic Books.

Pinker, Steven. 2007. *The Stuff of Thought.* New York: Viking.

Pinker, Steven. 2018. "We're told to fear robots. But why do we think they'll turn on us?" *Popular Science.* February 13, 2018. https://www.popsci.com/robot-uprising-enlightenment-now.

Porter, Jon. 2018. "Safari's suggested search results have been promoting conspiracies, lies, and misinformation." *The Verge.* September 26, 2018.

Preti, Maria Giulia, Thomas A. W. Bolton, and Dimitri Van De Ville. 2017. "The dynamic functional connectome: State-of-the-art and perspectives." *Neuroimage* 160: 41–54. https://doi.org/10.1016/j.neuroimage.2016.12.061.

Puig, Xavier, Kevin Ra, Marko Boben, Jiaman Li, Tingwu Wang, Sanja Fidler, and Antonio Torralba. 2018. "VirtualHome: Simulating household activities via programs." In *Computer Vision and Pattern Recognition.* https://arxiv.org/abs/1806.07011.

Pylyshyn, Xenon, ed. 1987. *The Robot's Dilemma: The Frame Problem in Artificial Intelligence.* Norwood, NJ: Ablex Pubs.

Quito, Anne. 2018. "Google's astounding new search tool will answer any question by reading thousands of books." *Quartz.* April 14, 2018. https://qz.com/1252664/talk-to-books-at-ted-2018-ray-kurzweil-unveils-googles-astounding-new-search-tool-will-answer-any-question-by-reading-thousands-of-books/.

Rahimian, Abtin, Ilya Lashuk, Shravan Veerapaneni, Aparna Chandramowlishwaran, Dhairya Malhotra, Logan Moon, Rahul Sampath, et al. 2010. "Petascale direct numerical simulation of blood flow on 200k cores and heterogeneous architectures." In *Supercomputing 2010*, 1–11. http://dx.doi.org/10.1109/SC.2010.42.

Rajpurkar, Pranav, Jian Zhang, Konstantin Lopyrev, and Percy Liang. 2016. "Squad: 100,000+ questions for machine comprehension of text." *arXiv preprint arXiv:1606.05250*. https:arxiv.org/abs/1606.05250.

Ramón y Cajal, Santiago. 1906. "The structure and connexions of neurons." Nobel Prize address. December 12, 1906. https://www.nobelprize.org/uploads/2018/06/cajal-lecture.pdf.

Rashkin, Hannah, Maarten Sap, Emily Allaway, Noah A. Smith, and Yejin Choi. 2018. "Event2Mind: Commonsense inference on events, intents, and reactions." *arXiv preprint arXiv:1805.06939*. https://arxiv.org/abs/1805.06939.

Rayner, Keith, Alexander Pollatsek, Jane Ashby, and Charles Clifton, Jr. 2012. *Psychology of Reading*. New York: Psychology Press.

Reddy, Siva, Danqi Chen, and Christopher D. Manning. 2018. "CoQA: A conversational question answering challenge." *arXiv preprint arXiv:1808.07042*. https://arxiv.org/abs/1808.07042.

Reece, Bryon. 2018. *The Fourth Age: Smart Robots, Conscious Computers, and the Future of Humanity*. New York: Atria Press.

Rips, Lance J. 1989. "Similarity, typicality, and categorization." In *Similarity and Analogical Reasoning*, edited by Stella Vosniadou and Andrew Ortony, 21–59. Cambridge: Cambridge University Press.

Rohrbach, Anna, Lisa Anne Hendricks, Kaylee Burns, Trevor Darrell, and Kate Saenko. 2018. "Object hallucination in image captioning." *arXiv preprint arXiv:1809.02156*. https://arxiv.org/abs/1809.02156.

Romm, Joe. 2018. "Top Toyota expert throws cold water on the driverless car hype." *ThinkProgress*. September 20, 2018. https://thinkprogress.org/top-toyota-expert-truly-driverless-cars-might-not-be-in-my-lifetime-0cca05ab19ff/.

Rosch, Eleanor H. 1973. "Natural categories." *Cognitive Psychology* 4(3): 328–350. https://doi.org/10.1016/0010-0285(73)90017-0.

Rosenblatt, Frank. 1958. "The perceptron: A probabilistic model for information storage and organization in the brain." *Psychological Review* 65(6): 386–408. http://psycnet.apa.org/record/1959-09865-001/.

Ross, Casey. 2018. "IBM's Watson supercomputer recommended 'unsafe and incorrect' cancer treatments, internal documents show." *STAT*, July 25, 2018. https://www.statnews.com/2018/07/25/ibm-watson-recommended-unsafe-incorrect-treatments/.

Ross, Lee. 1977. "The intuitive psychologist and his shortcomings: Distortions in the attribution process." *Advances in Experimental Social Psychology* 10: 173–220. https://doi.org/10.1016/S0065-2601(08)60357-3.

Roy, Abhimanyu, Jingyi Sun, Robert Mahoney, Loreto Alonzi, Stephen Adams, and Peter Beling. "Deep learning detecting fraud in credit card transactions." In *Systems and Information Engineering Design Symposium (SIEDS)*, 2018, 129–134. IEEE, 2018. https://doi.org/10.1109/SIEDS.2018.8374722.

Rumelhart, David E., Geoffrey E. Hinton, and Ronald J. Williams. 1986. "Learning representations by back-propagating errors." *Nature*. 323(6088): 533–536. https://doi.org/10.1038/323533a0.

Russell, Bertrand. 1948. *Human Knowledge: Its Scope and Limits*. New York: Simon and Schuster.

Russell, Bryan C., Antonio Torralba, Kevin P. Murphy, and William T. Freeman. 2008. "Labelme: A database and web-based tool for image annotation." *International Journal of Computer Vision,* 77(1–3): 157–173. http://www.cs.utsa.edu/~qitian/seminar/Spring08/03_28_08/LabelMe.pdf.

Russell, Stuart, and Peter Norvig, 2010. *Artificial Intelligence: A Modern Approach*. 3rd ed. Upper Saddle River, NJ: Pearson.

Ryan, V. 2001–2009. "History of Bridges: Iron and Steel." http://www.technologystudent.com/struct1/stlbrid1.htm Accessed by the authors, August 2018.

Sample, Ian. 2017. "Ban on killer robots urgently needed, say scientists." *The Guardian*. November 12, 2017. https://www.theguardian.com/science/2017/nov/13/ban-on-killer-robots-urgently-needed-say-scientists.

Sartre, Jean-Paul. 1957. "Existentialism is a humanism." Translated by Philip Mairet. In *Existentialism from Dostoevsky to Sartre,* edited by Walter Kaufmann, 287–311. New York: Meridian.

Schank, Roger, and Robert Abelson. 1977. *Scripts, Plans, Goals, and Understanding*. Hillsdale, NJ: Lawrence Erlbaum Associates.

Schoenick, Carissa, Peter Clark, Oyvind Tafjord, Peter Turney, and Oren Etzioni. 2016. "Moving beyond the Turing test with the Allen AI science challenge." *arXiv preprint arXiv:1604.04315*. https://arxiv.org/abs/1604.04315.

Schulz, Stefan, Boontawee Suntisrivaraporn, Franz Baader, and Martin Boeker. 2009. "SNOMED reaching its adolescence: Ontologists' and logicians' health check." *International Journal of Medical Informatics* 78: S86–S94. https://doi.org/10.1016/j.ijmedinf.2008.06.004.

Sciutto, Jim. 2018. "US intel warns of Russian threat to power grid and more."

*CNN*. July 24, 2018. https://www.cnn.com/videos/politics/2018/07/24/us-intel-warning-russia-cyberattack-threats-to-power-grid-sciutto-tsr-vpx.cnn/video/playlists/russia-hacking/.

Sculley, D. Gary Holt, Daniel Golovin, Eugene Davydov, Todd Phillips, Dietmar Ebner, Vinay Chaudhary, and Michael Young. 2014. "Machine learning: The high-interest credit card of technical debt." *SE4ML: Software Engineering 4 Machine Learning (NIPS 2014 Workshop)*. http://www.eecs.tufts.edu/~dsculley/papers/technical-debt.pdf.

Sejnowski, Terrence. 2018. *The Deep Learning Revolution*. Cambridge, MA: MIT Press.

Seven, Doug. 2014. "Knightmare: A DevOps cautionary tale." *Doug Seven* (blog). April 17, 2014. https://dougseven.com/2014/04/17/knightmare-a-devops-cautionary-tale/.

Shultz, Sarah, and Athena Vouloumanos. 2010. "Three-month-olds prefer speech to other naturally occurring signals." *Language Learning and Development* 6: 241–257. https://doi.org/10.1080/15475440903507830.

Silver, David. 2016. "AlphaGo." Invited talk, Intl. Joint Conf. on Artificial Intelligence. http://www0.cs.ucl.ac.uk/staff/d.silver/web/Resources_files/AlphaGo_IJCAI.pdf Accessed by the authors, December 26, 2018.

Silver, David, Aja Huang, Chris J. Maddison, Arthur Guez, Laurent Sifre, George Van Den Driessche, Julian Schrittwieser, et al. 2016. "Mastering the game of Go with deep neural networks and tree search." *Nature* 529(7587): 484–489. https://doi.org/10.1038/nature16961.

Silver, David, Julian Schrittwieser, Karen Simonyan, Ioannis Antonoglou, Aja Huang, Arthur Guez, Thomas Hubert, et al. 2017. "Mastering the game of Go without human knowledge." *Nature* 550(7676): 354–359. https://doi.org/10.1038/nature24270.

Silver, David, et al. 2018. "A general reinforcement learning algorithm that masters chess, shogi, and Go through self-play." *Science* 362(6419): 1140–1144. http://doi.org/10.1126/science.aar6404.

Simon, Herbert. 1965. *The Shape of Automation for Men and Management*. New York: Harper and Row.

Simonite, Tom. 2019. "Google and Microsoft warn that AI may do dumb things." *WIRED,* February 11, 2019. https://www.wired.com/story/google-microsoft-warn-ai-may-do-dumb-things/.

Singh, Push, Thomas Lin, Erik T. Mueller, Grace Lim, Travell Perkins, and Wan Li Zhu. 2002. "Open Mind Common Sense: Knowledge acquisition from the general public." In *OTM Confederated International Confer-*

*ences "On the Move to Meaningful Internet Systems,"* 1223–1237. Berlin: Springer. https://doi.org/10.1007/3-540-36124-3_77.

Skinner, B. F. 1938. *The Behavior of Organisms*. New York: D. Appleton-Century.

Skinner, B. F. 1957. *Verbal Behavior*. New York: Appleton-Century-Crofts.

Smith, Gary. 2018. *The AI Delusion*. Oxford: Oxford University Press.

Soares, Nate, Benja Fallenstein, Stuart Armstrong, and Eliezer Yudkowsky. 2015. "Corrigibility." In *Workshops at the Twenty-Ninth Conference of the American Association for Artificial Intelligence (AAAI)*. https://www.aaai.org/ocs/index.php/WS/AAAIW15/paper/viewPaper/10124.

Solon, Olivia. 2016. "Roomba creator responds to reports of 'poopocalypse': 'We see this a lot.'" *The Guardian*. August 15, 2016. https://www.theguardian.com/technology/2016/aug/15/roomba-robot-vacuum-poopocalypse-facebook-post.

Souyris, Jean, Virginie Wiels, David Delmas, and Hervé Delseny. 2009. "Formal verification of avionics software products." In *International Symposium on Formal Methods,* 532–546. Berlin, Heidelberg: Springer. https://www.cs.unc.edu/~anderson/teach/comp790/papers/Souyris.

Spelke, Elizabeth. 1994. "Initial knowledge: six suggestions." *Cognition*. 50(1–3): 431–445. https://doi.org/10.1016/0010-0277(94)90039-6.

Sperber, Dan, and Deirdre Wilson. 1986. *Relevance: Communication and Cognition*. Cambridge, MA: Harvard University Press.

Srivastava, Nitish, Geoffrey Hinton, Alex Krizhevsky, Ilya Sutskever, and Ruslan Salakhutdinov. 2014. "Dropout: A simple way to prevent neural networks from overfitting." *Journal of Machine Learning Research* 15(1): 1929–1958. http://www.jmlr.org/papers/volume15/srivastava14a/srivastava14a.pdf.

Statt, Nick. 2018. "Google now says controversial AI voice calling system will identify itself to humans." *The Verge*. May 10, 2018. https://www.theverge.com/2018/5/10/17342414/google-duplex-ai-assistant-voice-calling-identify-itself-update.

Sternberg, Robert J. 1985. *Beyond IQ: A Triarchic Theory of Intelligence*. Cambridge: Cambridge University Press.

Stewart, Jack. 2018. "Why Tesla's Autopilot can't see a stopped firetruck." *WIRED*. August 27, 2018. https://www.wired.com/story/tesla-autopilot-why-crash-radar/.

Sweeney, Latanya. 2013. "Discrimination in online ad delivery." *Queue* 11(3): 10. https://arxiv.org/abs/1301.6822.

Swinford, Echo. 2006. *Fixing PowerPoint Annoyances.* Sebastopol, CA: O'Reilly Media.

Tegmark, Max. 2017. *Life 3.0: Being Human in the Age of Artificial Intelligence.* New York: Alfred A. Knopf.

Thompson, Clive. 2016. "To make AI more human, teach it to chitchat." *WIRED.* January 25, 2016. https://www.wired.com/2016/01/clive-thompson-12/.

Thrun, Sebastian. 2007. "Simultaneous localization and mapping." In *Robotics and Cognitive Approaches to Spatial Mapping,* edited by Margaret E. Jeffries and Wai-Kiang Yeap, 13–41. Berlin, Heidelberg: Springer. https://link.springer.com/chapter/10.1007/978-3-540-75388-9_3.

Tomayko, James. 1998. *Computers in Spaceflight: The NASA Experience.* NASA Contractor Report 182505. https://archive.org/details/nasa_techdoc_19880069935.

Tullis, Paul. 2018. "The world economy runs on GPS. It needs a backup plan." *Bloomberg BusinessWeek.* July 25, 2018. https://www.bloomberg.com/news/features/2018-07-25/the-world-economy-runs-on-gps-it-needs-a-backup-plan.

Turing, Alan. 1950. "Computing machines and intelligence." *Mind* 59:433–460.

Turkle, Sherry. 2017. "Why these friendly robots can't be good friends to our kids." *The Washington Post,* December 7, 2017. https://www.washingtonpost.com/outlook/why-these-friendly-robots-cant-be-good-friends-to-our-kids/2017/12/07/bce1eaea-d54f-11e7-b62d-d9345ced896d_story.html.

Ulanoff, Lance. 2002. "World Meet Roomba." *PC World.* September 17, 2002. https://www.pcmag.com/article2/0,2817,538687,00.asp.

Vanderbilt, Tom. 2012. "Let the robot drive: The autonomous car of the future is here." *WIRED.* January 20, 2012. https://www.wired.com/2012/01/ff_autonomouscars/.

van Harmelen, Frank, Vladimir Lifschitz, and Bruce Porter, eds. 2008. *The Handbook of Knowledge Representation.* Amsterdam: Elsevier.

Van Horn, Grant, and Pietro Perona. 2017. "The devil is in the tails: Fine-grained classification in the wild." *arXiv preprint arXiv:1709.01450.* https://arxiv.org/abs/1709.01450.

Vaswani, Ashish, Noam Shazeer, Niki Parmar, Jakob Uszkoreit, Llion Jones, Aidan N. Gomez, Łukasz Kaiser, and Illia Polosukhin. 2017. "Attention is all you need." In *Advances in Neural Information Processing Systems,* 5998–6008. http://papers.nips.cc/paper/7181-attention-is-all-you-need.

Veloso, Manuela M., Joydeep Biswas, Brian Coltin, and Stephanie Rosenthal. 2015. "CoBots: Robust symbiotic autonomous mobile service robots." *Proceedings of the Intl. Joint Conf. on Artificial Intelligence 2015:* 4423–4428. https://www.aaai.org/ocs/index.php/IJCAI/IJCAI15/paper/view Paper/10890.

Venugopal, Ashish, Jakob Uszkoreit, David Talbot, Franz J. Och, and Juri Ganitkevitch. 2011. "Watermarking the outputs of structured prediction with an application in statistical machine translation." *Proceedings of the Conference on Empirical Methods in Natural Language Processing:* 1363–1372. https://dl.acm.org/citation.cfm?id=2145576.

Vigen, Tyler. 2015. *Spurious Correlations*. New York: Hachette Books.

Vincent, James. 2018a. "Google 'fixed' its racist algorithm by removing gorillas from its image-labeling tech." *The Verge*. January 12, 2018. https://www.theverge.com/2018/1/12/16882408/google-racist-gorillas-photo-recognition-algorithm-ai.

Vincent, James. 2018b. "IBM hopes to fight bias in facial recognition with new diverse dataset." *The Verge*. June 27, 2018. https://www.theverge.com/2018/6/27/17509400/facial-recognition-bias-ibm-data-training.

Vincent, James. 2018c. "OpenAI's Dota 2 defeat is still a win for artificial intelligence." *The Verge*. August 28, 2018. https://www.theverge.com/2018/8/28/17787610/openai-dota-2-bots-ai-lost-international-reinforcement-learning.

Vincent, James. 2018d. "Google and Harvard team up to use deep learning to predict earthquake aftershocks." *The Verge*. August 30, 2018. https://www.theverge.com/2018/8/30/17799356/ai-predict-earthquake-aftershocks-google-harvard.

Vinyals, Oriol. 2019. "AlphaStar: Mastering the real-time strategy game StarCraft II." Talk given at New York University, March 12, 2019.

Vinyals, Oriol, Alexander Toshev, Samy Bengio, and Dumitru Erhan. 2015. "Show and tell: A neural image caption generator." In *Proceedings of the IEEE Conference on Computer Vision and Pattern Recognition,* 3156–3164. https://ieeexplore.ieee.org/stamp/stamp.jsp?arnumber=7505636.

Vondrick, Carl, Aditya Khosla, Tomasz Malisiewicz, and Antonio Torralba. 2012. "Inverting and visualizing features for object detection." *arXiv preprint arXiv:1212.2278*. https://arxiv.org/abs/1212.2278.

Wallach, Wendell, and Colin Allen. 2010. *Moral Machines: Teaching Robots Right from Wrong*. Oxford: Oxford University Press.

Walsh, Toby. 2018. *Machines That Think: The Future of Artificial Intelligence.* Amherst, NY: Prometheus Books.

Wang, Alex, Amapreet Singh, Julian Michael, Felix Hill, Omer Levy, and Samuel R. Bowman. 2018. "GLUE: A multi-task benchmark and analysis platform for natural language understanding." *arXiv preprint arXiv:1804.07461*. https://arxiv.org/abs/1804.07461.

Watson, John B. 1930. *Behaviorism*. New York: W. W. Norton.

Weizenbaum, Joseph. 1965. *Computer Power and Human Reason.* Cambridge, MA: MIT Press.

Weizenbaum, Joseph. 1966. "ELIZA—a computer program for the study of natural language communication between man and machine." *Communications of the ACM* 9(1): 36–45.

Weston, Jason, Sumit Chopra, and Antoine Bordes. 2015. "Memory networks." *Int. Conf. on Learning Representations,* 2015. https://arxiv.org/abs/1410.3916.

Wiggers, Kyle. 2018. "Geoffrey Hinton and Demis Hassabis: AGI is nowhere close to being a reality." *VentureBeat*. December 17, 2018. https://venturebeat.com/2018/12/17/geoffrey-hinton-and-demis-hassabis-agi-is-nowhere-close-to-being-a-reality/.

Wikipedia. "Back propagation." https://en.wikipedia.org/wiki/Backpropagation. Accessed by authors, December 2018.

Wikipedia. "Driver verifier." https://en.wikipedia.org/wiki/Driver_Verifier. Accessed by authors, December 2018.

Wikipedia. List of countries by traffic-related death rate. Accessed by authors, December 2018. https://en.wikipedia.org/wiki/List_of_countries_by_traffic-related_death_rate.

Wikipedia. "OODA Loop." https://en.wikipedia.org/wiki/OODA_loop. Accessed by authors, December 2018.

Wilde, Oscar. 1891. "The Soul of Man Under Socialism." *Fortnightly Review*. February 1891.

Wilder, Laura Ingalls. 1933. *Farmer Boy.* New York: Harper and Brothers.

Willow Garage. 2010. "Beer me, Robot." *Willow Garage* (blog). http://www.willowgarage.com/blog/2010/07/06/beer-me-robot.

Wilson, Benjamin, Judy Hoffman, and Jamie Morgenstern. 2019. "Predictive inequity in object detection." *arXiv preprint arXiv:1902.11017*. https://arxiv.org/abs/1902.11097.

Wilson, Chris. 2011. "Lube job: Should Google associate Rick Santorum's name with anal sex?" *Slate*. July 1, 2011. http://www.slate.com/articles/technology/webhead/2011/07/lube_job.html.

Wilson, Dennis G., Sylvain Cussat-Blanc, Hervé Luga, and Julian F. Miller. 2018. "Evolving simple programs for playing Atari games." *arXiv preprint arXiv:1806.05695*. https://arxiv.org/abs/1806.05695.

Wissner-Gross, Alexander. 2014. "A new equation for intelligence." TEDx-BeaconStreet talk. November 2013. https://www.ted.com/talks/alex_wissner_gross_a_new_equation_for_intelligence.

Wissner-Gross, Alexander, and Cameron Freer. 2013. "Causal entropic forces." *Physical Review Letters* 110(16): 168702. https://doi.org/10.1103/PhysRevLett.110.168702.

Witten, Ian, and Eibe Frank. 2000. *Data Mining: Practical Machine Learning Tools and Techniques with Java Implementation*. San Mateo, CA: Morgan Kaufmann.

Wittgenstein, Ludwig. 1953. *Philosophical Investigations*. London: Blackwell.

WolframAlpha Press Center. 2009. "Wolfram|Alpha officially launched." https://www.wolframalpha.com/media/pressreleases/wolframalpha-launch.html. As of December 27, 2018, this web page is no longer functional, but it has been saved in the Internet Archive at https://web.archive.org/web/20110512075300/https://www.wolframalpha.com/media/pressreleases/wolframalpha-launch.html.

Woods, William A. 1975. "What's in a link: Foundations for semantic networks." In *Representation and Understanding*, edited by Daniel Bobrow and Allan Collins, 35–82. New York: Academic Press.

Yampolskiy, Roman. 2016. *Artificial Intelligence: A Futuristic Approach*. Boca Raton, FL: CRC Press.

Yudkowsky, Eliezer. 2011. "Artificial intelligence as a positive and negative factor in global risk." In *Global Catastrophic Risks*, edited by Nick Bostrom and Milan Cirkovic. Oxford: Oxford University Press.

Zadeh, Lotfi. 1987. "Commonsense and fuzzy logic." In *The Knowledge Frontier: Essays in the Representation of Knowledge*, edited by Nick Cercone and Gordon McCalla, 103–136. New York: Springer Verlag.

Zhang, Baobao, and Allan Dafoe. 2019. *Artificial Intelligence: American Attitudes and Trends*. Center for the Governance of AI, Future of Humanity Institute, University of Oxford, January 2019. https://governanceai.github.io/US-Public-Opinion-Report-Jan-2019/high-level-machine-intelligence.html.

Zhang, Yu, William Chan, and Navdeep Jaitly. 2017. "Very deep convolutional networks for end-to-end speech recognition." In *IEEE International Conference on Acoustics, Speech and Signal Processing,* 4845–4849. https://doi.org/10.1109/ICASSP.2017.7953077.

Zhou, Li, Jianfeng Gao, Di Li, Heung-Yeung Shum. 2018. "The design and implementation of XiaoIce, an empathetic social chatbot." *arXiv preprint 1812.08989*. https://arxiv.org/abs/1812.08989.

Zito, Salena. 2016. "Taking Trump seriously, not literally." *The Atlantic*. September 23, 2016. https://www.theatlantic.com/politics/archive/2016/09/trump-makes-his-case-in-pittsburgh/501335/.

Zogfarharifard, Ellie. 2016. "AI will solve the world's 'hardest problems': Google chairman, Eric Schmidt, says robots can tackle overpopulation and climate change." *Daily Mail*. January 12, 2016. https://www.dailymail.co.uk/sciencetech/article-3395958/AI-solve-world-s-hardest-problems-Google-chairman-Eric-Schmidt-says-robots-tackle-overpopulation-climate-change.html.

# Index

Page numbers in **bold face** indicate an extended discussion of the term.

# ILLUSTRATION CREDITS

Page 22: O. Vinyals, A. Toshev, S. Bengio, and D. Erhan, under Creative Commons 3.0
Page 23: O. Vinyals, A. Toshev, S. Bengio, and D. Erhan, under Creative Commons 3.0
Page 31: Maayan Harel
Page 37: Maayan Harel
Page 45: Gary Marcus
Page 46: Ernest Davis
Page 49: K. Fukushima, I. Hayashi, and J. Léveillé
Page 58: Anish Athalye, Logan Engstrom, Andrew Ilyas, Kevin Kwok
Page 59: (top) Anish Athalye, Logan Engstrom, Andrew Ilyas, Kevin Kwok; (middle, bottom left, bottom right) T. B. Brown, D. Mané, A. Roy, M. Abadi, and J. Gilmer
Page 60: (top) I. Evtimov, K. Eykholt, E. Fernandes, T. Kohno, B. Li, A. Prakash, and D. Song; (bottom) Michael A. Alcorn, Qi Li, Zhitao Gong, Chengfei Wang, Long Mai, Wei-Shinn Ku, and Anh Nguyen
Page 66: Randall Munroe
Page 75: Maayan Harel
Page 81: Steve Luck
Page 82: Maayan Harel
Page 96: IEEE Spectrum
Page 97: Maayan Harel
Page 131: Ernest Davis
Page 132: Gary Lupyan and Andy Clark
Page 133: (top) Gary Lupyan and Andy Clark; (bottom) L. Carmichael, H. P. Hogan, and A. A. Walter
Page 134: L. Carmichael, H. P. Hogan, and A. A. Walter
Page 135: (all) Matthew G. Bisanz, under Creative Commons 3.0
Page 137: Maayan Harel
Page 140: Tyler Vigen
Page 148: © by ACM, Inc. September 2015, created by Peter Crowther
Page 152: C. Havasi, J. Pustejovsky, R. Speer, and H. Lieberman
Page 156: Maayan Harel
Page 158: Maayan Harel
Page 159: Maayan Harel
Page 165: (left) Emj, under Creative Commons 3.0; (right) Jonasek22, under Creative Commons 3.0
Page 189: Maayan Harel
Page 193: Maayan Harel

## ABOUT THE AUTHORS

Gary Marcus is a scientist, best-selling author, and entrepreneur. He is founder and CEO of Robust.AI and was founder and CEO of Geometric Intelligence, a machine-learning company acquired by Uber in 2016, and is the author of five books, including *Kluge, The Birth of the Mind*, and the *New York Times* best seller *Guitar Zero*.

Ernest Davis is Professor of Computer Science at the Courant Institute of Mathematical Science, New York University. He is one of the world's leading experts on commonsense reasoning for artificial intelligence, and the author of four books, including *Representations of Commonsense Knowledge* and *Verses for the Information Age*.

## A NOTE ON THE TYPE

The text of this book was set in Sabon, a typeface designed by Jan Tschichold (1902–1974), the well-known German typographer. Based loosely on the original designs by Claude Garamond (ca. 1480–1561), Sabon is unique in that it was explicitly designed for hot-metal composition on both the Monotype and Linotype machines as well as for filmsetting. Designed in 1966 in Frankfurt, Sabon was named for the famous Lyons punch cutter Jacques Sabon, who is thought to have brought some of Garamond's matrices to Frankfurt.

*Composed by North Market Street Graphics,*
*Lancaster, Pennsylvania*

*Printed and bound by Berryville Graphics,*
*Berryville, Virginia*

*Designed by Betty Lew*